SHE management systems

for small to medium-sized enterprises

SHE management systems

for small to medium-sized enterprises

Anders Jacobsson

AJ Risk Engineering AB

The procedures in this book have been based on good industrial practice and are produced in good faith, but this does not imply the acceptance of any legal liability or responsibility whatsoever, by the European Process Safety Centre or its officers, the authors or by the Institution of Chemical Engineers as publisher, for the consequences of its use or misuse in any particular circumstances. This disclaimer shall have effect only to the extent permitted by any applicable law.

The legal requirements concerning safety, occupational health and environmental protection in different countries vary widely. It is therefore not possible to produce a set of procedures which can meet all of these requirements.

Whilst the procedures contained in this book may be used for guidance, it remains the responsibility of those in charge of the operation concerned to satisfy themselves:

- that they are adequate and appropriate to control the hazards involved;
- that they fully meet the legal requirements of the country in which the operations are being undertaken.

Published by
Institution of Chemical Engineers (IChemE),
Davis Building,
165–189 Railway Terrace,
Rugby, Warwickshire CV21 3HQ, UK
IChemE is a Registered Charity

© 2000 European Process Safety Centre

ISBN 0 85295 433 6

Front cover photographs:
Centre: a fully sealed NIMEX mixing vessel (courtesy of NIMEX Limited, Cranfield, UK).
Bottom right: 18 MW test furnace at Hamworthy Combustion Engineering Limited's Advanced Technology Centre
(courtesy of Hamworthy Combustion Engineering Limited).

Printed in the United Kingdom by Page Bros, Norwich, UK

Foreword

The production and use of chemicals frequently involves the handling of hazardous materials. Experience built up by chemical manufacturers over many years has established methods of working which, when correctly applied, reduce the risks involved to tolerable levels.

Over the last decade major companies in the process industries worldwide have found that the safety of their operations can be improved even further if the individual procedures are integrated into a Safety Management System (or a Safety, Health and Environment Management System).

In 1994 a sub-committee of the European Process Safety Centre (EPSC), drawing on the experience of major companies in the process industries from across Europe, produced a book on the topic, *Safety Management Systems: Sharing experiences in process safety*.

The importance of safety management systems has been further recognized by regulatory authorities and incorporated as one of the key elements in the Seveso II Directive.

Between 1996 and 1997 a committee of the Association of Swedish Chemical Industries (Kemikontoret) worked on the production of a set of procedures which could be used by smaller concerns and those who are involved in the use of chemicals. Following discussions between Kemikontoret and EPSC, it was agreed that the work would be translated into English and published by EPSC. The main author of the original work, Anders Jacobsson, kindly agreed to undertake this and has now completed the translation resulting in this book.

EPSC is pleased to publish the procedures contained in this guide in the expectation that they will help small to medium-sized enterprises to improve their performance. EPSC gratefully acknowledges the contribution of Kemikontoret in supporting the original work and by agreeing to its being shared with a wider audience.

Robin Turney
Technical Director
European Process Safety Centre
October 2000

Contents

Foreword   iii

METHODOLOGY

Summary   1

Why is a management system needed?   3

Safety, Health and Environment (SHE) as an integrated system 4

The relationship between SHE and quality   5

Structure of the SHE management system   6

Developing the SHE management system   8

Implementing the SHE management system   10

Keeping the SHE management system running   11

Independent audit and certification/registration   12

How to use the procedures   13

Model company for the procedures   14

PROCEDURES

0. SHE management system, general   17

1. SHE policy   19

2. Organization and management   20
2.1 Organization   20
2.2 SHE objectives and action plans   23
2.3 Management review   24

3. Legislation and permits   26
3.1 SHE legislation   26
3.2 SHE permits   28

4. General SHE information   30
4.1 SHE effects/impacts   30
4.2 SHE reporting   32

5. Personnel, training and communication   34
5.1 Health care   34
5.2 Training   36

5.3	Internal communication	40
5.4	External communication	41

6.	**Purchasing**	**43**
6.1	Purchasing	43

7.	**Operations and maintenance**	**46**
7.1	General site rules	46
7.2	Instructions	48
7.3	Plant integrity and maintenance	50
7.4	Work permits	55
7.5	Contractors	64
7.6	Handling of chemical products	66
7.7	Handling of wastes	68
7.8	Soil and groundwater protection	71
7.9	Non-process hazards	73
7.10	Management of change	74
7.11	Process safety (SHE) information	79

8.	**Engineering and projects**	**81**
8.1	Engineering standards	81
8.2	Investment projects	82
8.3	Manufacturing methods/R&D work/scale-up	86
8.4	Technology sale	89

9.	**Auditing and inspection**	**90**
9.1	SHE audits	90
9.2	Risk analysis	92
9.3	SHE rounds	94
9.4	Occupational health	96
9.5	Environmental control	98

10.	**Emergency response**	**101**
10.1	Emergency response	101
10.2	Fire protection	103

11.	**Accidents, incidents and disturbances**	**107**
11.1	Actions in the case of accidents, incidents and disturbances	107
11.2	Investigation and reporting of accidents, incidents and disturbances	108

12.	**Transport**	**110**
12.1	Transport	110

13.	**Product control**	**112**
13.1	Product control	112

Further reading	**114**
Appendix 1 – Reference key to ISO 14001: 1996 (EMAS)	**115**
Appendix 2 – Reference key to BS 8800: 1996	**117**

Methodology

Summary

Objectives of this guide

This guide aims to outline a simple way of achieving an integrated Safety, Health and Environment (SHE) management system.

A SHE management system is a structured approach to managing an operation and encourages continuous improvement whilst ensuring that the company complies with the relevant SHE legislation. This guide can form the basis for applying a formal SHE standard and becoming certified, for example to ISO 14001.

This guide's primary emphasis lies on the procedures for a management system which are presented in a form suitable for a 'model company'. A process for developing a SHE management system is also covered briefly.

The objective of the guide is to allow organizations to build efficient SHE management systems, specific to their own requirements, based on the methodology and the procedures used by the model company. Every company will of course have to make adjustments to the example illustrated in this guide, but a large proportion of the contents will be generally applicable.

A simple, yet flexible, model for a management system can easily be developed and the company can be prepared for any external requirements — such as new legislation or market pressure.

Whilst some companies will be able to carry out much of this work with their own resources, others will gain from the use of specialist consultants or others. However, be aware that the work will require effort, resources and commitment.

Scope

The management system is made up of Safety, Health and Environment as an integrated concept. The Responsible Care initiative, promoted and supported by chemical manufacturers, has been used as a basis for the work.

Any management system should ideally cover the basic requirements of respective standards including:

- ISO 14001 (Environmental management systems — Specification with guidance for use);
- EMAS (Eco Management and Audit Scheme regarding environmental management systems);
- BS 8800 (British Standard 'Guide to Occupational Health and Safety management Systems');
- ICC's (International Chamber of Commerce) 16-point programme;
- Scandinavian Regulations for Internal Control of the working environment.

A SHE management system needs to regulate the factors that can influence any part of the operation that can give rise to significant SHE effects. The system should therefore be based on the results obtained from an initial assessment of the SHE hazards and risks within an organization.

Model

The system described in this guide is a model based on activities common to most companies. It includes all of the elements which are found in more formal SHE systems, but it has also been made flexible enough to adjust to the needs of individual companies whilst trying to be compatible with the possible future requirements of legislation and standards.

The management system illustrated in this guide comprises 40 procedures, each regulating an activity common in companies handling chemicals. They have been classified into 13 groups, ranging from general to very specific procedures.

To illustrate the SHE management system outlined in this guide, the procedures have been written using a model company, Chemicals Handling Inc., a company with about 50 employees.

To optimize the benefit of the system, the procedures all use the following standardized structure:

- objectives;
- scope;
- principles and methods;
- responsibilities;
- references.

Goal setting

In expressing specific requirements in the procedures included in this guide, there is a suggestion of what could be a reasonable goal to aim for. Some companies will already exceed this level, while for others significant effort may be required to achieve it. Companies are of course free to choose their own goals, with legislatory requirements setting a minimum level.

It is important that companies choose a target goal which is achievable. If the goal is initially set too high, this will have a negative impact on the whole process. It is better to begin with a somewhat lower goal which is achievable, and then successively increase it in achievable increments.

Auditing

To ensure that the management system fulfils its basic requirements and is effective in practice, it must be regularly audited. It should be noted that the structures of a SHE management system and an effective auditing method are very similar. Sources of information on auditing techniques are included in the 'Further Reading' section of this guide (see page 114).

Certification/registration

One objective of this guide is that the SHE management systems based on the recommended procedures contained in it should be able to fulfil all the requirements and lead to certification/registration with the systems such as ISO 14001 and EMAS.

Integration with other management systems

To obtain the most benefit from a SHE management system it should be integrated with existing company management systems.

Commitment of company management

To be successful a SHE management system needs full visible commitment from senior management, as with any other management system. The senior management must give full support through resources and active participation in the development and implementation of the system.

Why is a management system needed?

Managing a company's SHE policy in a professional way should be as natural as managing production, marketing, personnel, economy and finance. Formal systems and tools are needed for this, as with any of the company's other management systems.

Safety, Health and Environment issues are often regulated in a company by a number of different loosely connected instructions and procedures and a general and comprehensive policy. Some companies have built up a culture within one or two of the SHE areas through long traditions. Most companies will find that their SHE performance can be improved by bringing all the relevant procedures together into one integrated SHE management system.

Poor management is often the root cause of accidents

Analysis and investigations show that weakness in the management of operations is one of the most common causes of accidents. It is often found that there is a lack of procedures, or poor procedures, poor adherence and poor follow-up of the procedures by the management.

Principles and commitment

An effective SHE management system should clearly show the management's basic principles and its commitment to SHE. A formal SHE management system contributes to illustrating a company's approach to structuring SHE issues.

Driving forces

The interest in management systems has increased considerably with various market forces and some regulatory requirements driving this development. Within the environmental area, standards such as the ISO 14000 series and EMAS, have made many companies decide to develop and certify/register management systems. In the recent EU directive, Seveso II 96/82/EU, there are requirements for certain companies to have a safety management system. This directive was implemented by the EU member states in 1999.

Increased credibility

The implementation of a SHE management system can lead to better relationships, an improved reputation and credibility both within the company (among employees and union organizations) and externally (among neighbours, community, the media, competitors and authorities).

Further advantages

In addition to the fact that a SHE management system meets regulatory requirements and makes the effort devoted to SHE issues more effective, there will also be:

- more efficient production and maintenance as a result of fewer operational disturbances, fewer releases, less absenteeism, etc;
- more effective project activities and simplified start-up through considering SHE questions at an early stage;
- better relationships with customers, neighbours and suppliers.

Properly implemented, a SHE management system will prove to be cost-effective.

Safety, Health and Environment (SHE) as an integrated system

Safety, Health and Environment issues can be seen as a highly integrated concept. Separate management systems for Safety, Health or Environment therefore require many common or identical parts — for example:

- a policy;
- responsibility for tasks;
- competence and training of personnel;
- operating instructions/controls, measurement and documentation;
- auditing.

The same general principles are required to manage the operations so it is therefore logical to integrate Safety, Health and Environment into a common management system. Managing the SHE issues in this way is more effective because there will be significantly less documentation to keep track of, to update, to train people in the use of, and for the employees to follow.

There are also financial gains to be made in comparison to having separate systems for environmental management, occupational health, Responsible Care, etc.

An integrated system may make it more difficult for an authority or a certifying body to audit an integrated system compared to tailor-made systems for every individual application. This should be seen as being more than out-weighed by the benefits that an integrated system can bring to an organization. There are clear signs that some authorities are interested in seeing integrated systems within companies, which has resulted in authorities making more integrated inspections.

Integration of Safety, Health and Environment					
Health				**Environment**	
	Safety				
Individual's health needs	Work-related illnesses	Accidents at work	Risk of major accidents	Acute environmental impact	Chronic environmental impact

The relationship between SHE and quality

Whether SHE and quality should be two separate management systems (with some relationships), or totally integrated with each other, is often debated.

This guide aims to provide examples of procedures such that companies can develop the contents in specific SHE procedures. Whether these should then be totally separate, constitute a separate SHE management system or form part of a larger management system can be decided by the organization later. If a company already has a quality management system, for instance ISO 9001/2, it could consider integrating the SHE procedures into the existing system. However, this can lead to the risk of losing focus on SHE issues.

When a totally separate SHE management system is selected, many of the basic principles and the technical design of the system can be developed from the existing quality management system. Further use can be made of an existing quality management system by cross reference from the SHE management system. It may also be appropriate to add certain SHE aspects to the quality system — for example, under existing maintenance and calibration procedures. However, regardless of which approach is taken, it is important to avoid duplication of work between the systems.

Structure of the SHE management system

The system is built around the normal activities of the company

The SHE management system illustrated in this guide starts with an analysis of the normal activities of a company. The activities which are of importance for SHE issues are thereafter regulated in the form of SHE procedures.

This system therefore deviates in its structure from, for example, ISO 9001/2 and ISO 14001, because the latter starts from more general activities such as 'planning', 'implementation', 'follow-up' and so on. The system in this guide covers these generic points under each activity.

The SHE management system should be based on the significant SHE aspects, which are identified during an initial assessment.

Other formal systems

This guide has been devised such that demands made by other systems such as ISO 14001, EMAS, BS 8800, the ICC 16-point programme and Responsible Care have been built into the system. The examples supplied in this guide include a number of reference keys to facilitate a check against these other systems (see Appendices 1 and 2 on pages 115–118). These should be of assistance in a certification process.

General structure

A SHE management system should be structured according to the following hierarchical principles:
1. Policy
2. Procedures
3. Instructions

The policy gives the overall view and objectives within the area. The procedures outline what should be done and generally when, where, how and by whom. In some cases the procedures are complemented with detailed instructions on how and by whom the activities shall be performed. Sometimes it can be beneficial to distribute the system in a manual or to assemble all the procedures in a handbook.

The model within this guide contains proposals for producing a policy and a large number of example procedures.

Procedures

The guide contains almost 40 SHE procedures to regulate important activities from a SHE point of view.

Each procedure has a standardized structure with fixed headings as follows:
- objectives;
- scope;
- principles and methods;
- responsibility;
- references.

The procedures should be seen as examples.

Once a company has assessed the Safety, Health and Environmental effects of its operation it needs to assess whether the procedures described here are adequate to control the hazards and meet legislative requirements.

In many cases the text is relatively comprehensive to show the scope of the suggestions, target goals, etc. In a finalized procedure this background text could be omitted.

Document control

It is important that all relevant documents within the system are covered by effective document control, primarily to ensure that only relevant documents are available where the work is performed. This can be achieved by heading all procedures with a registration number, the revision number, the validity date, who is the author and who is responsible for the procedure and its distribution.

A correctly functioning document control system is important in achieving an effective SHE management system.

These guidelines do not detail document control systems, allowing each company to choose its own model. The procedure headings serve as examples of the most important aspects to be controlled.

Model company

In order to illustrate the SHE management system and clarify the procedures, they have been written for a fictitious company called 'Chemicals Handling Inc., a small company with a relatively standard organizational structure.

The procedures emphasize that the direct responsibility for SHE lies within the line organization. In the model company there is also a staff function responsible for SHE issues. This could be one person, such as a safety engineer or environmental co-ordinator. For the purposes of this guide, this person is assumed to be the manager of the SHE management system itself.

Within Chemicals Handling Inc. there is a SHE committee which is an extended version of the safety committee. For those companies which are not formally required to have a safety committee, the participation of personnel could be achieved in other ways.

The aim of this guide is that every company, whatever its size or structure, should be able to adjust the procedures to its own activities.

Flexible system

The structure of the SHE management system means that it should be easy to develop the system whenever additional requirements arise. For example, the company may consider that there is another activity to be regulated, there may be new requirements in legislation, or standards, or from some stakeholder. By working in this way the company can adjust the system to its optimum and be well prepared for any new requirements.

Developing the SHE management system

For a SHE management system to be successfully implemented and become well established there are a number of principles that must be followed at the development stage. Some of the most important are:

- to sell the concept to the top management, including the company managing director, and ask for active support and involvement;
- to appoint one person who works as the 'champion' for the project;
- to involve the whole organization in the work;
- to work through a project and/or a steering group;
- to make a time schedule with targets and sub-targets;
- to build the system on any existing procedures, such as the company's quality management system.

The involvement and support of management

The success of a SHE management system depends on the involvement of the company's senior management. The organization will act based upon the priority it believes the management has given the project.

The managing director/site manager must speak on behalf of the system and show that he or she supports the basic ideas. The management must show a sustainable involvement; it is not enough just to start the process. It is therefore crucial for the management to guarantee the build-up, implementation and operation of the SHE management system, in terms of monetary and personnel resources. Monitoring the development and implementation of the SHE management system is an obvious task for the company management group in their regular group meetings. The status and performance of the SHE work should also be reported at the company board level.

Involvement of the whole organization

Involving the whole organization means, among other things, that personnel are actively involved in developing the procedures. The personnel must have the opportunity to say how they can work with the SHE issues. The system will be difficult to implement if this is not done during the development phase.

Project organization and 'champion'

Careful organization is needed to carry out a SHE management project effectively.

Overseeing the project organization should be a project manager. This person is the project champion, and the task will require significant time resources during the development, training and implementation phases.

A project group should be formed with a representative mix of people. This group will contribute to the development of the procedures within the system and provide an anchor for the procedures and the target goals within their respective departments.

Personnel resources will be needed from all departments. Groups consisting of foremen, operators, technicians, union representatives, etc. may be needed as bodies for consultation during the work. A proposed structure for the project within Chemicals Handling Inc. could be:

Project Director:
Managing Director/Site Manager

Project Manager (champion):
SHE Manager

Project group:
SHE Manager (chairman)
Production Manager
Foreman for operations
Marketing/Sales Manager
Personnel/Administration/Purchasing Manager
Main Safety Representative
(The project group would need to meet approximately once every 2–4 weeks.)

Work/consultation groups from the departments:
Foremen, operators, technicians, safety representatives, etc.
(Working groups can provide a continuous checking process through weekly meetings.)

Consider outside assistance

Whilst much of the work described in this guide can be carried out by an organization's own staff, careful consideration should be given to the use of outside assistance in the form of consultants and others. This can bring a fresh view to the project, provides experience and expertise drawn from other organizations and, most importantly, the time to devote to the project.

Outside assistance will reduce the likelihood of mistakes or major factors being overlooked and is likely to increase the value of the investment being made by the company.

Time schedule

The development and implementation of a SHE management system must be conducted within a time frame with suitable resources made available. The time taken from the decision to start developing the system through to implementation can range from six to 24 months. This is dependent on how much of the procedures and other material is already available and the resources that the company is prepared to commit to the project. Some of the activities requiring a significant investment of time include:

- the initial SHE assessment;
- the development and construction of the system;
- training and participation of personnel from the whole organization;
- detailed regulation and instructions at department level (if this is considered necessary).

The use of the guide's model for both the SHE management system and procedures can shorten the time considerably.

Before considering undertaking any certification or registration, the system should have been running for a certain period of time — approximately six months.

Building on existing systems

A SHE management system should build upon existing foundations, which the company has within the SHE area. Sound written or verbal rules and instructions which are already applied should be used as important corner stones in the new system. Use should be made as far as possible of existing material.

A first step in developing a complete SHE management system is to make a survey of existing operations, procedures and documentation in the company. This first step can be called an 'initial assessment', with the results compared with the suggested contents of a SHE management system in this guide. A first estimation can then be obtained as to which areas and to what extent work has to be carried out to arrive at a comprehensive SHE management system. An assessment may be required of any company-specific and important SHE influencing activities which require additional procedures to those proposed in this guide. This can be judged when the initial SHE assessment has been carried out (according to a special procedure contained in this guide).

Paper or electronic systems

If a company already has well-developed procedures available electronically then it would be advantageous to also put the SHE management system in such a form. One advantage of a paper-free system is that it enables far more efficient document control and reduces the likelihood of employees using older/outdated documents. However, if the company does not make use of computerized systems, the SHE management system in paper form can prove equally effective. If a company is planning to develop and introduce electronic systems, it is inadvisable to do this at the same time as introducing a SHE management system.

Implementing the SHE management system

The implementation of the SHE management system is a critical part of the project. This can be the case even if employees have been involved in developing the procedures. Many questions can arise at this point — for example:

- how will the system be received?
- how well will it be followed?
- what training is needed?
- what control and follow-up is needed?

Therefore a lot of attention and pre-planning should be given to the implementation stage. The project management and the company management should allocate extra resources during the actual implementation of the management system and a renewed drive for motivating the personnel is advisable.

Stepwise implementation

In larger companies the SHE management system can be introduced on one unit first as a pilot case, before it is introduced in the whole company. In smaller companies it is normally more effective to introduce it across the whole organization at the same time. However, in such cases it is advisable to introduce the system in steps rather than all at once, and utilize the experiences from the procedures introduced first into the procedures which are still to be implemented.

If it is decided to implement the procedures stepwise there are various strategies available — ranging from starting with the simplest procedures to starting with the most challenging ones. It is probably advisable to select some procedures which have the potential to be well received — not too controversial and yet not too self-evident.

Follow-up

The project manager must closely monitor how the first procedures are received by employees and how they work in practice. This can be achieved through interviews with employees at all levels, reviews of how employees are utilizing the documents, etc. The follow-up should continue until the process is established within the line organization.

The line organization's responsibility

Line management, from site manager down to first line supervisors, are responsible for ensuring that the procedures are adhered to. The production manager has special responsibilities in this chain.

Training

Before or during the implementation of the management system, training of all personnel is needed, and should include:

- managers;
- operators;
- technicians;
- contractors and transporters.

It is important to allocate sufficient resources for training, with the formal part of the training varying in length from hours to days. There will also be a lot of informal training and discussions within the organization.

Keeping the SHE management system running

The introduction of a set of procedures is clearly not enough in itself to keep a SHE management system in operation. A number of additional steps are required with the overall goal of 'continual improvement'. The diagram below illustrates a common model for ensuring continuous improvement — a loop starting with the SHE policy, and passing through planning, implementation and operation, checking and corrective action, and management review, to provide continual improvement.

It is only when the whole 'loop' is completed that the system can be considered to be in operation and mature enough for a possible external audit.

For the SHE management system to be successful the following activities are required:
- continuous measurement of improvements;
- periodic control in the form of internal and external auditing;
- continuous training and motivation of personnel.

Responsibility for managing the system

A champion, such as the manager of the company's SHE function, must continue to be responsible for supervising, controlling and developing the system.

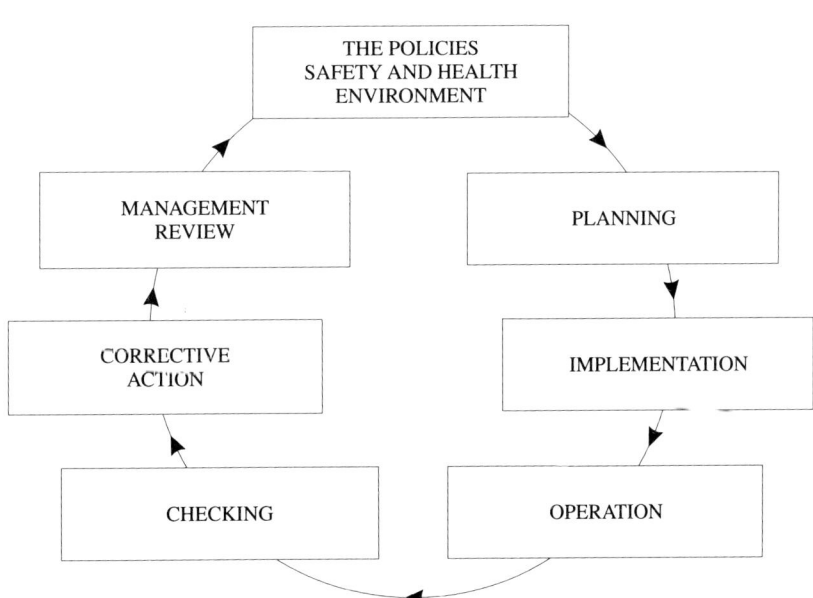

SHE management system model.

Independent audit and certification/registration

There are several reasons for a company to certify its SHE management system. Through certification or registration, the company is acknowledged to have fulfilled the requirements of a certain system/standard.

The certificate can simplify communication with external stakeholders or with a growing number of customers now requiring their suppliers to be certified according to ISO 14001, OHSAS 18001 (BS 8800) or registered according to EMAS.

Some authorities have indicated that they would sometimes be prepared to accept that certified/registered companies have achieved a good basic standard and, in the future, fewer detailed inspections will be made of such companies.

It is important to emphasize that certification or registration should not be the main reason for a company to develop a SHE management system. There must be a genuine desire to improve in the area of SHE by introducing such a management system.

There are many good reasons for allowing an independent party to audit the system, and companies can contract a certified auditor to carry out an audit. This has usually been to obtain certification according to ISO 14001 or registration according to EMAS, and recently also certification according to OHSAS 18001.

Some of the most important conditions for achieving certification or registration are:
- the system shall be in operation;
- the system shall fulfil all the requirements according to the standard and the legislation;
- the system documentation shall be in good order — this applies to both controlling documents and recording documents.

In order to ensure that all detailed requirements in the standard or the directive have been included in the SHE management system, it could be appropriate to make a reference key between the standard and the company system. Such a key would also greatly assist the work of an auditor. See Appendices 1 and 2 on pages 115–118 for examples.

How to use the procedures

The procedures within this guide are intended as an aid for companies developing their own SHE management systems. They aim to cover the majority of the requirements of companies that handle chemicals.

The procedures give advice on:

- which procedures may be needed to regulate the enterprise concerned;
- what needs to be regulated within each activity (scope, methods, responsibility);
- which target goal is appropriate;
- how the structure of a management system can be developed;
- what the structure of individual procedures might look like.

The procedures within this guide have purposely been written to contain sufficient detail to provide background information. Each company must decide how to apply the procedures to suit its own activities, either deleting or retaining the additional text.

The company should choose a target goal for its own company procedures by subtracting from or adding to those parts where a target goal is expressed. Considering the actual conditions and goals of the company, a decision can then be reached on the extent to which a proposed procedure should be utilized. Additional information on setting goals can be found in the publications given in the 'Further reading' section of this guide (see page 114).

The majority of the procedures within this guide are written so that they provide a reasonable target for most companies to achieve or exceed as a long-term policy.

To illustrate, an example is given below concerning 'reporting and follow-up of incidents and disturbances (SHE)'. As a comparison there is also a statement of the requirements which would be assessed as being less than satisfactory.

Reasonable target/goal
'Well developed, written system exists. The personnel of the unit/department report immediately all (also minor) incidents, near-misses and disturbances on a regular basis.'

Less than satisfactory
'Reporting and follow-up system exists. Reporting is only done for serious incidents and disturbances.'

Irrespective of which target goal is chosen, it is important that the procedure expresses clearly the scope and objectives. It is also important to state who within the company is responsible for which activity in the procedure.

Some procedures are very detailed, such as the procedure for 'modifications'. This is because such procedures reflect areas which when neglected have been shown to cause accidents.

In the following procedures there are many using the term 'shall'. In some cases the final procedure of the company may contain a similar formulation using 'shall', but in many cases this 'shall' is a request on the company to specify how a requirement will be fulfilled.

Model company for the procedures

In order to use completed SHE procedures as illustrations in this guide, they have been prepared for a fictitious model company, Chemicals Handling Inc. The diagram below shows the structure of the model company.

Chemicals Handling Inc. is relatively small, with about 50 employees. It is relatively independent and there are no overall corporate procedures for Safety, Health and Environment (SHE).

The company manufactures a limited number of products. Some of the operations require the use of shift work, normally two-shift but sometimes three-shift. There is a small maintenance workforce and contractors are also used. The company receives raw materials and chemicals, and exports its products, through road transport.

A limited amount of research and development work on new and modified products is carried out on site. As a support for the operational activities there are in-house resources for SHE, quality and design/engineering.

In addition to the technical functions there are departments handling marketing/sales, personnel/administration and purchasing.

The model company is a small company producing chemicals. However, the SHE procedures which have been developed can also be applied to both larger and smaller enterprises in other businesses — for example, companies in the energy sector, mechanical engineering industry, food industry, pulp and paper industry, and so on.

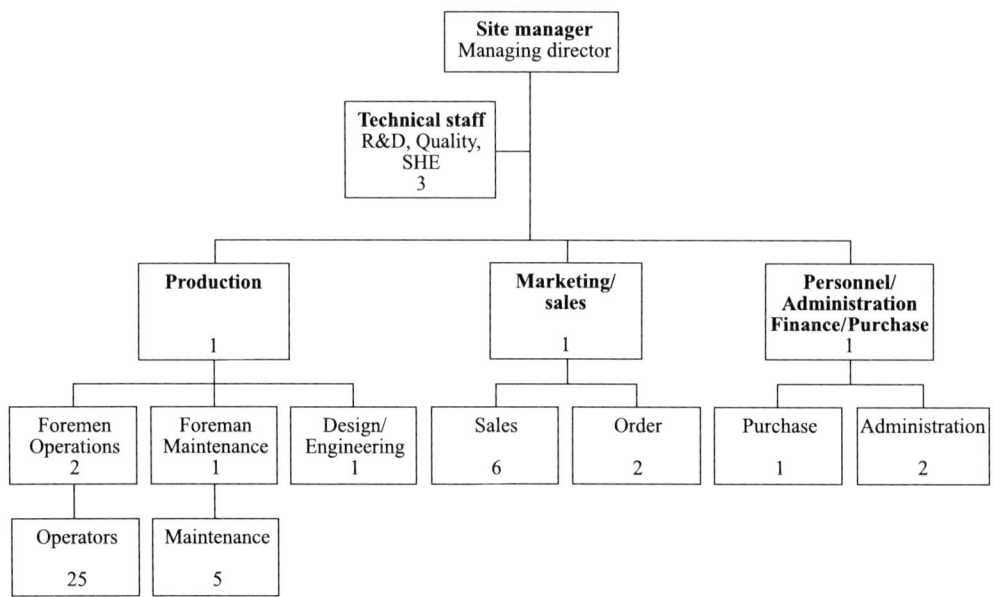

Structure of the model company 'Chemicals Handling Inc.'

Procedures

Chemicals Handling Inc., SHE Management Procedures **SHE Management System, General**				
Reg. No: SHE-0		Revision: 1	Valid from:	Page 1 of 2
Approved by:		Date:		Issued by:
Distribution:				

Objective	This basic procedure describes the Safety/Health/Environment (SHE) management system for Chemicals Handling Inc. The objective of the SHE management system is to give a structure to and comprehensive control of the important SHE aspects of the company's operation. The system shall define an ambition level which the company considers adequate for its business. As a minimum, the requirements of legislation shall under all circumstances be fulfilled. Further objectives of the SHE management system are: ● more efficient production and maintenance with fewer operating disturbances, releases, less absenteeism, etc; ● more efficient project management and smoother start-up by incorporating SHE requirements at an early stage; ● improved customer relations; ● improved relations and increased reputation within the company among employees and union organizations and externally with the authorities, the public, the community, the media, the competitors, etc.
Scope	The SHE management system applies to the whole enterprise of Chemicals Handling Inc. The system shall control all the activities which could affect safety, health or the environment.
Principles and methods	Chemicals Handling Inc. has chosen to build the SHE management system based on the normal activities of the company. **Structure of the system** Chemicals Handling Inc.'s commitment to SHE issues is set out in the company SHE policy, which is an integral part of the SHE management system. In order to fulfil this policy the company has regulated all those activities which could have some SHE effect – either under normal conditions or during abnormal events – in the form of SHE procedures. The SHE procedures are thus subordinate to the SHE policy and are an integral part of the SHE management system. In some cases more detailed regulation is needed. This is set out in the form of instructions. In most cases the SHE aspects are integrated into the instructions used for the normal operation of the company. **Contents of procedures** The general structure of the procedures is: ● objective; ● scope (regarding activity, contents and geographically); ● principles and methods; ● responsibility; ● references. The procedures state what shall be done, the requirements and the ambition level which should be achieved. They also define which person(s) in the organization is responsible for ensuring that the procedures are followed. The procedures also indicate how to attain the goals and when and where certain activities should be performed.

Chemicals Handling Inc., SHE Management Procedures **SHE Management System, General**				
Reg. No: SHE-0		Revision: 1	Valid from:	Page 2 of 2
Principles and methods (cont'd)	**Document control** All documents which are part of the formal SHE management system should be easy to identify and locate. All the formal SHE management system shall be subject to document control. This ensures that the system is kept updated, and that all those concerned have the latest revision of every document. A comprehensive list of documents shall be kept, stating name of document, filing place, filing time and responsible person. **SHE management for continual improvement** The Chemicals Handling Inc. SHE management system is designed in such a way that the goal of 'continual improvement' shall be attained in the work. Based on the SHE policy the following steps should be taken in turn: • planning; • implementation and operation; • control and corrective action; • management review; in a continuous loop, by which continual improvement is achieved. The system is constructed in such a way that it could easily be supplemented if so required. The system should be easy to audit.			
Responsibility	Line management is responsible for the management of SHE aspects of all the operations under its control in accordance with the SHE policy and SHE management system. Overall responsibility rests with the Managing Director.			
References	This procedure refers to the following SHE procedures: ALL			

Chemicals Handling Inc., SHE Management Procedures **SHE Policy**				
Reg. No: SHE-1		Revision: 1	Valid from:	Page 1 of 1
Approved by:		Date:		Issued by:
Distribution:				

Objective	The objective of this SHE policy is to form a basis for Chemicals Handling Inc.'s way of working within the areas of Safety, Health and Environment. Based on the policy, the company has worked out a number of SHE procedures in order to regulate in detail how various activities within its business shall be performed in order to fulfil the requirements of the policy.
Scope	The SHE policy applies to all the activities within Chemicals Handling Inc.
Commitments	<ul><li>Safety, Health and Environment are the highest priorities for Chemicals Handling Inc. They are a natural part of the management of the company. SHE aspects shall always be considered as part of the decision-making process.</li><li>The care for man and environment shall always have priority in the business of the company. The company aims to do business without causing significant damage within or outside the company.</li><li>The management of Chemicals Handling Inc. has the primary responsibility for SHE.</li><li>All employees have a personal responsibility for SHE.</li><li>Chemicals Handling Inc. shall, as a minimum, always fulfil the relevant legislation and other official rules. The company aims to be better in most areas.</li><li>The employees of Chemicals Handling Inc. shall have good knowledge of SHE issues and be committed to high standards of performance.</li><li>Chemicals Handling Inc. shall effectively use raw material, energy and other natural resources.</li><li>The workplace and the processes shall be designed, and the work carried out, in such a way that risks for people and the environment from these activities are as low as reasonably practicable.</li><li>Chemicals Handling Inc. shall strive to use the best available technology in terms of SHE. Where this is not possible the risk for man and environment shall be minimized by safety devices, personal protection equipment, good work procedures, training, etc.</li><li>Chemicals Handling Inc. shall regularly review and analyse the SHE aspects of its activities and strive for continual improvement.</li><li>The SHE work shall be characterized by minimizing risks and damaging effects, and not by restoration after damage.</li><li>Incidents and accidents shall immediately be reported and investigated in order to prevent a recurrence. During disturbances, safety, health and environment shall be given priority. Reporting events shall always be considered as a positive action, whilst omitting to report will be seen as a negative action.</li><li>Safety, health and environmental issues shall be discussed in an open and positive atmosphere within Chemicals Handling Inc. Openness shall also be characterized by sharing information and co-operating with authorities, the public and other external stakeholders.</li><li>Chemicals Handling Inc. shall give its customers SHE information regarding its products and general activities.</li><li>The company shall ensure that contractors, co-operation partners, etc. will follow the same SHE principles and rules which apply to Chemicals Handling Inc. while working within the boundaries of the company site and activities.</li><li>Chemicals Handling Inc. shall regularly review and ensure that the activities are performed according to this policy and all SHE procedures. This shall be carried out by SHE audits and by internal reviews.</li></ul>
Responsibility	The company management is responsible for observing this policy and updating it when necessary.
	Signed: .... (Managing Director, Chemicals Handling Inc.) Date: ....

Chemicals Handling Inc., SHE Management Procedures **Procedure for Organization**				
Reg. No: SHE-2.1	Revision: 1	Valid from:		Page 1 of 3
Approved by:	Date:		Issued by:	
Distribution:				

Objective	Having clear responsibilities is absolutely essential if an organization is to function well. The objective of this procedure is to clarify the principles for the distribution of work tasks with the appropriate responsibility and authority within Chemicals Handling Inc. and to describe the administrative system which regulates this.
Scope	This procedure applies to the whole organization and all the job positions and bodies within Chemicals Handling Inc. which can affect SHE issues.
Principles and methods	The organization of Chemicals Handling Inc. is presented in Attachment 1. **A basic principle for the responsibility for safety/health/environment in the company is that it is a line responsibility.** The regulation of responsibilities is primarily done in two ways. **1) Delegation from the managing director/site manager to department managers** The managing director has, in special delegation documents, delegated job tasks including the associated responsibility for occupational health and the environment to the department managers. In certain cases a further delegation to the next hierarchical level may be made. These delegation documents describe the role of the persons who have been delegated responsibility, and the scope and definition of this responsibility. The document shall be signed by both parties. **2) Job descriptions** All job positions within the company shall have a job description. The description defines the contents, responsibility and authority of the position. Responsibility and authority regarding safety, health and environment especially shall be clearly spelled out. The job descriptions shall be revised in case of changes or when otherwise needed. In addition to this a general review and update shall be carried out at defined intervals. Each department manager is responsible for the correctness of the job descriptions and for the review of them. Overall responsibility for the job descriptions system lies with the personnel department. **Responsibility and roles outside the line organization** *Chemicals Handling Inc.'s consultation committee* Chemicals Handling Inc. has a consultation committee, mainly consisting of the management and employee representatives. The consultation committee shall have SHE issues as a standing point on the agenda. *Management SHE representative* The company management has appointed the manager of the SHE function as its representative in SHE matters. This person shall have the necessary competence in SHE questions. The person has the responsibility on behalf of the management to implement and administer the SHE management system. Resources and authority for this are defined in the job description.

Chemicals Handling Inc., SHE Management Procedures **Procedure for Organization**				
Reg. No: SHE-2.1		Revision: 1	Valid from:	Page 2 of 3

Principles and methods (cont'd)	*SHE committee*
	
	 Within the company there is a SHE committee which is an enlargement of the safety committee. The composition of the SHE committee is defined in Attachment 1. The main tasks of the committee are to agree goals, to monitor progress towards these goals and to ensure that a system of two-way communication is in place which will enable all employees to participate in their achievement.
	
	     Decisions on policy and other overall decisions on SHE are taken by the company management. The SHE committee shall be consulted by management on the basis for such decisions and any other affecting SHE.
	
	 SHE function
	 There is a staff function for SHE issues in the organization. This function has the functional responsibility to supervise the work in the plant and the departments to ensure it is carried out in a professional way according to the policies, rules and instructions of Chemicals Handling Inc. Naturally the law and other regulatory requirements are the primary prerequisites for the business. The SHE function has, besides its supervising function, a supporting and recommending duty to the line organization. It is responsible for following the development of legislation, etc., supporting the line organization with this knowledge and following up how this is implemented in the various activities. It shall also follow up and report SHE issues to the authorities.
	
	     The SHE function has no formal right of veto against the line organization. In the case of a disagreement, the company management decides.
	
	 Emergency response organization
	 Under certain circumstances connected with an emergency situation, a special emergency response organization comes into existence and takes over the responsibility of the facilities. This organization shall be able to function at all times—day and night all year round. The emergency response organization is described in more detail in the company emergency plan.
	
	 Co-ordination responsibility for contractors
	 The responsibility for the co-ordination of work which is fully or partly carried out by contractors lies with the plant owner, i.e. Chemicals Handling Inc. The responsibilities shall be regulated in the contract with the respective contractor.
	
	 Changes of the organization
	 Changes of the organization shall be treated according to the SHE procedure for MANAGEMENT OF CHANGE, SHE-7.10.

Responsibility	The site manager/managing director has overall responsibility to maintain clear conditions for responsibility and the associated authority in the organization according to this procedure.
References	This procedure refers to the following SHE procedures: SHE POLICY MANAGEMENT OF CHANGE SHE LEGISLATION EMERGENCY RESPONSE CONTRACTORS

Chemicals Handling Inc., SHE Management Procedures
Procedure for Organization
Attachment 1 — Company organization

Reg. No: SHE-2.1	Revision: 1	Valid from:	Page 3 of 3
			Attachment 1

SHE committee
- Site manager (chairman)
- SHE manager (secretary)
- Production manager
- Operator
- Foreman
- Union representative
- Medical representative

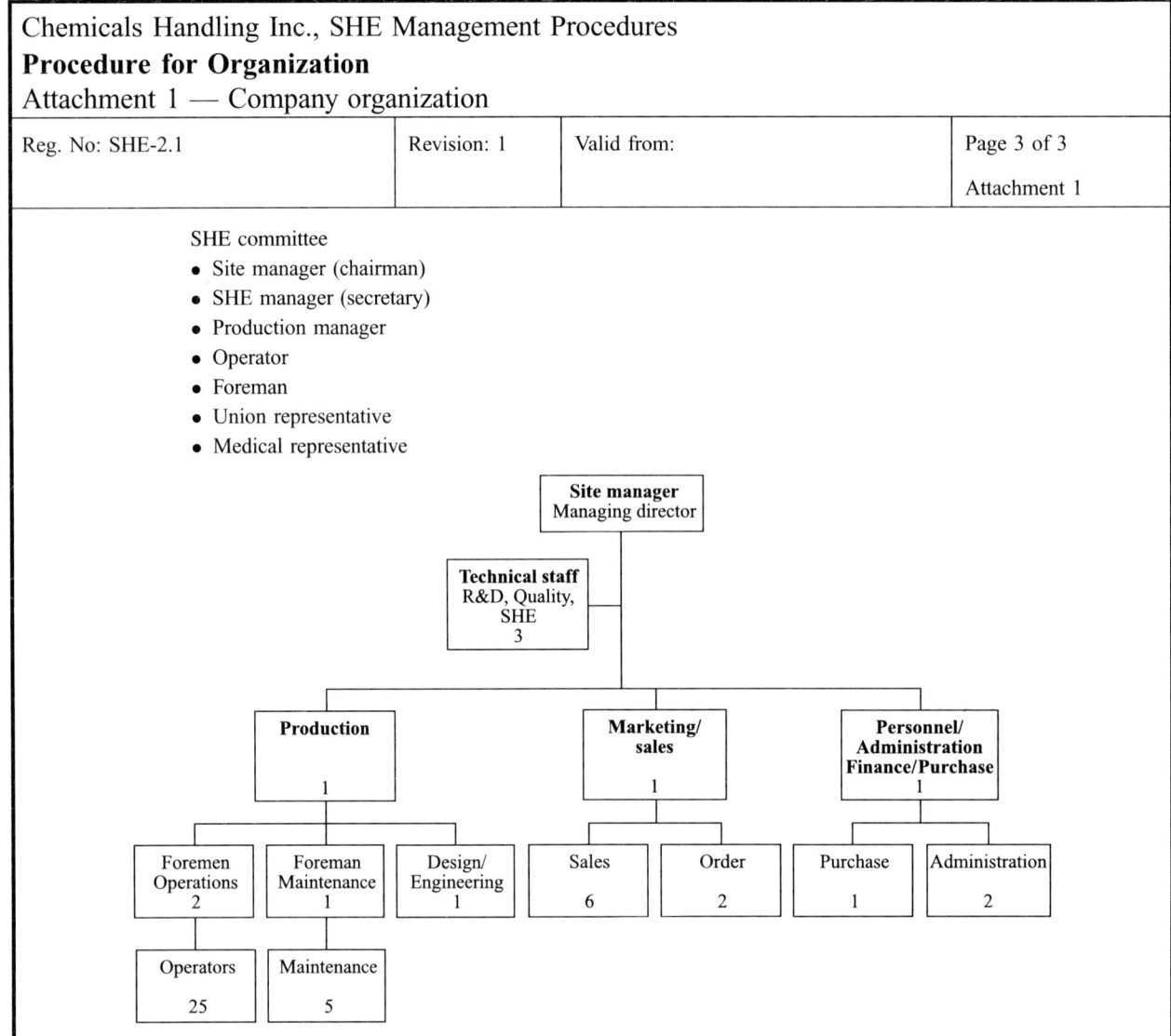

Chemicals Handling Inc., SHE Management Procedures **Procedure for SHE Objectives and Action Plans**					
Reg. No: SHE-2.2		Revision: 1	Valid from:		Page 1 of 1
Approved by:		Date:		Issued by:	
Distribution:					
Objective	The objective of this procedure is to achieve continual improvement within the SHE area.				
Scope	This procedure applies to all activities within Chemicals Handling Inc. The company shall establish objectives/ targets and action plans for safety/health/environment on the following levels: company level and department level.				
Principles and methods	The objectives/targets shall be based on the SHE policy and on the actual SHE effects which are caused by the company's activities (which have been identified according to the procedure for SHE EFFECTS/IMPACTS, (SHE-4.1). Targets for SHE shall be set annually in connection with the normal budget process. The targets shall be decided in the budget process by the same party which approves the financial budget. **Type of targets** Chemicals Handling Inc. shall establish both general objectives/targets and detailed targets. The targets shall normally be formulated so that they are specific, measurable, challenging, realistic and understandable. They should preferably be quantitative but qualitative targets may also be used. Examples of targets can be: • decreasing the number of accidents; • increasing the reporting of incidents and near-misses; • carrying out a project or an activity (e.g., occupational health measurement, inventory of environmental releases, decrease of waste amount, specific health controls); • carrying out a specific SHE training activity; • improving the neighbour's view of the company's SHE work. **Action programme** Every target shall have a matching action programme. This should show activities, sub-targets and final targets. For each activity a responsible person should be appointed. Every activity shall be time planned and when required costs shall be calculated and approved. **Follow-up and reporting** Follow-up and reporting shall be done in writing in the quarterly and yearly SHE reports and also regularly at departmental meetings, management group meetings and board meetings. **Documentation** Targets and action programmes shall be in writing.				
Responsibility	The site manager/managing director, and the department manager are responsible for formulating, defining and implementing the targets. The departmental targets shall also be overseen by the SHE committee and approved by the management group.				
References	This procedure refers to the following SHE procedures: SHE POLICY SHE EFFECTS/IMPACTS SHE LEGISLATION				

Chemicals Handling Inc., SHE Management Procedures **Procedure for Management Review**				
Reg. No: SHE-2.3	Revision: 1	Valid from:		Page 1 of 2
Approved by:	Date:		Issued by:	
Distribution:				
Objective	The objective of this procedure is to ensure that the management of Chemicals Handling Inc. undertakes a periodic review of the status of SHE in the company, which can be used as a basis for continual improvement.			
Scope	This procedure applies to all activities within Chemicals Handling Inc.			
Principles and methods	The management of Chemicals Handling Inc. shall regularly control the status of SHE aspects of the company's activities. The review shall mainly determine that: • Chemicals Handling Inc. fulfils the spirit of its policy; • SHE targets which have been defined and agreed are attained; • the management system is effective and the SHE policy relevant. **Scope and review** The review shall comprise (but not be limited to): • the yearly department SHE review (according to the procedure for SHE AUDITS, SHE-9.1); • results of any independent audits which have been carried out; • reporting of accidents, near-misses and disturbances; • environmental reporting; • reports on health issues and occupational hygiene; • ensuring that the relevant legislation is followed; • interviews with key personnel. **Frequency** The management team shall perform this review once per year in a special meeting. In between these yearly reviews the management shall keep abreast of any developments by informal reviews. These reviews should be performed immediately before defining next year's SHE targets. **Documentation** The result of the management review shall be documented. **Information** The findings of the review shall be communicated to the organization and to the board. **Follow-up of results** The review may conclude that the existing procedures must be adhered to more firmly. This should be done during the normal course of operation. The review may also lead to the development of new or modified SHE procedures, review and update of the SHE policy or the change of other controlling instruments if deemed necessary. The management review shall be the basis for setting the targets and activity plans for the next year.			

Chemicals Handling Inc., SHE Management Procedures **Management Review**			
Reg. No: SHE-2.3	Revision: 1	Valid from:	Page 2 of 2
Responsibility	The site manager/managing director is responsible for the management review.		
References	This procedure refers to the following SHE procedures: SHE POLICY SHE REPORTING ORGANIZATION SHE AUDITS SHE LEGISLATION SHE ROUNDS SHE PERMITS ACCIDENTS, INCIDENTS AND DISTURBANCES SHE EFFECTS/IMPACTS		

Chemicals Handling Inc., SHE Management Procedures **Procedure for SHE Legislation**				
Reg. No: SHE-3.1		Revision: 1	Valid from:	Page 1 of 2
Approved by:		Date:	Issued by:	
Distribution:				

Objective	The objective of this procedure is to create a system, to enable the company to keep track of and apply all relevant legislation within the SHE area.
Scope	This procedure applies to all areas and activities within Chemicals Handling Inc.
Principles and methods	Chemicals Handling Inc. shall follow all relevant legislation as a basic requirement. This means that the following areas of legislation (not necessarily limited to these) are applicable: ● environment; ● chemical products; ● flammable and explosive products; ● occupational health and safety; ● rescue service; ● transport of dangerous goods; ● electrical equipment; ● planning and building. (The list will vary from country to country.) **Documentation** A valid version of the relevant legislation should be available at the company. It is advisable to subscribe to the most relevant legislation, if possible. **Register** A register of valid legislation documents (laws, regulations, directives, etc.), which the company considers applicable to its activities shall be kept, indicating for each document where the original is kept and who is responsible for it (according to sample in Attachment 1). **Actions due to new or modified legislative requirements** When there are new or modified requirements to legislation, the SHE function, in co-operation with line managers, shall identify the activities in the company which could be affected by these changes. The line manager shall make an action programme for the relevant ones. 　　Once per year, the SHE function shall check that all relevant legislation is followed. The follow-up (in writing) shall be reported to the SHE committee and the management group.
Responsibility	It is the responsibility of the site manager/managing director (at the company level) and the department managers (the departmental level) to follow all the legislation. 　　The SHE function is responsible for keeping the relevant legislation documents updated. It is also the responsibility of the SHE function to review, interpret and follow changes and news in the area, and when these occur report and advise the line managers.
References	This procedure refers to the following SHE procedures: SHE POLICY

Chemicals Handling Inc., SHE Management Procedures
Procedure for SHE Legislation
Attachment 1 — Example of a register of legislation relevant to Chemicals Handling Inc.'s activities

| Reg. No: SHE-3.1 | Revision: 1 | Valid from: | Page 2 of 2 |
| | | | Attachment 1 |

Legislation	Filing place	Responsible	Comments
Occupational health and safety 1	SHE function	Department manager	
Occupational health and safety 2	SHE function	Department manager	
	SHE function	Department manager	
	Respective department		
Environmental protection	SHE function	Department manager	
Flammable and explosive material	SHE function	Department manager	
Chemical products			
Transport of dangerous goods			
Rescue service			

Chemicals Handling Inc., SHE Management Procedures **Procedure for SHE Permits**				
Reg. No: SHE-3.2		Revision: 1	Valid from:	Page 1 of 2
Approved by:		Date:		Issued by:
Distribution:				

Objective	The objective of this procedure is to create a simple system to: • keep all relevant permits available and in good order; • keep permits updated; • keep an overview of existing conditions and rules.
Scope	This procedure applies to all activities within Chemicals Handling Inc.
Principles and methods	The following principles are valid: • all activities which need a permit shall have a valid permit; • Chemicals Handling Inc. shall have the original versions of all necessary permits assembled in one specific filing place; • all permits are regulated in a series of SHE procedures for fulfilling the requirements and conditions. **Valid permits** Chemicals Handling Inc.'s permits are summarized in Attachment 1. **Updating** All permits shall be kept updated. Updating work shall be initialized well in advance, so that a new valid permit is always available when an old one runs out. All permits shall be reviewed and controlled by a responsible person at least once each year. **Notification** In cases where there is a legislative requirement to notify the relevant authority, the notification shall be documented and the document shall be filed.
Responsibility	The SHE function is responsible for keeping a register of relevant permits and looking after their validity.
References	This procedure refers to the following SHE procedures: SHE POLICY

Chemicals Handling Inc., SHE Management Procedures
Procedure for SHE Permits
Attachment 1 — Example of a summary of permits for Chemicals Handling Inc.

Reg. No: SHE-3.2	Revision: 1	Valid from:	Page 2 of 2
			Attachment 1

Permit according to legislation	Valid until	Filing place	Responsible	Comments
Concession according to environmental legislation	31/12/2001	SHE function	Department manager	
Permit to handle flammable and explosive material	31/12/2001	SHE function	Department manager	
Permit to handle carcinogenic substances				

Chemicals Handling Inc., SHE Management Procedures **Procedure for SHE Effects/Impacts**				
Reg. No: SHE-4.1		Revision: 1	Valid from:	Page 1 of 2
Approved by:		Date:		Issued by:
Distribution:				

Objective	This procedure shall ensure that the company makes a basic assessment and review of the status of its safety, health and environment issues. This basic assessment shall then be the starting point within the SHE area for: • procedures and instructions; • training; • targets and action plans. Risk assessments according to this procedure shall also fulfil the requirements of relevant legislation.
Scope	This procedure applies to all activities within Chemicals Handling Inc., including R&D activities, production, storing, marketing and distribution. The procedure comprises the effects of activities on safety, health and environment in the following aspects: 1) Effects at the site (employees and environment). 2) Effects on humans and the environment when producing, for example, the necessary raw materials for site activities. 3) Effects on humans and the environment whilst distributing and using the company products, and also whilst reusing, recycling the products or taking care of wastes.

Principles and methods	All units shall make an assessment of the SHE effects which the company activities cause or could cause. This shall deal with either continuous effects or effects from a sudden event.
	
	Contents
	
	Effects at the site
	The effects from normal activities (inclusive of start-up, shutdown, maintenance, etc.) shall be identified and documented in this section, as should abnormal situations (breakdown, sudden releases, fire, explosion, etc.). The effects of earlier activities which could be significant today or in the future shall also be documented.
	  Areas requiring treatment and the actual levels to be quantified are:
	
	• risks from unhealthy working conditions or accidents at the workplace (releases to the working environment of gas, liquid, dust, noise, etc., machine risks, etc.);
	• releases to the atmosphere (potential or actual);
	• releases to water or sewer systems;
	• wastes, especially toxic or dangerous wastes;
	• ground contamination;
	• 'release' of waste heat, noise, smell, dust, vibration, etc.
	
	  In most cases, existing investigations and measurements can be used, but new measurements, calculations or reviews may be needed in certain cases. Notable concentrations or amounts concerning the working environment shall be noted.
	  In this assessment, consideration should be made of historic 'near-misses' and accidents within the SHE area.
	  A general assessment of the usage of ground, water, fuels and other energy as well as other natural resources shall be made.
	  Simple material and energy balances for important parameters should be made.

Chemicals Handling Inc., SHE Management Procedures **Procedure for SHE Effects/Impacts**			
Reg. No: SHE-4.1	Revision: 1	Valid from:	Page 2 of 2

Principles and methods (cont'd)	In addition to the inventory of potential 'releases', an audit of the administrative procedures in the SHE area should be carried out in order to make a decision on whether the organization can efficiently control the SHE aspects. Such an audit is carried out according to the SHE procedure for SHE AUDITS, SHE-9.1. *Environmental impact of associated operations* As well as analysing the SHE consequences from the activities at the site, the units/departments shall carry out simplified analyses of the most important SHE aspects in the chain up to the point where the units start utilizing the raw materials and utilities (point 2 above) and then for the products or the semi-manufactured products leaving the units, including final disposal or recycle or reuse (point 3 above). The analysis can be made from making enquiries of the suppliers and the customers. The opportunities for the organization to influence the SHE situation regarding choice of raw materials, manufacturing processes and use of and disposal of the products shall be included in the analysis. The assessment can be documented in a similar way as effects at the site. *Treatment of abnormal events* Risk analyses regarding safety, health and environment shall be conducted for assessing the SHE risks from deviations from normal procedures. Methods for this are described in the SHE procedure for RISK ANALYSIS, SHE-9.2. A Preliminary Hazard Analysis shall be carried out as the lowest level for this analysis. Such risk analyses shall form part of the total assessment of the SHE effects of the company activities. **Updating** This inventory and assessment shall be updated once every three years or more frequently if necessary. The inventory shall be a living basis for the company's activities. **Special assessment** A special assessment of significant SHE effects shall be made. Basic facts in the procedure for RISK ANALYSIS, SHE-9.2 could be used as an assessment model. **Documentation** All assessments referred to above shall be documented and made available in central files. All significant SHE effects arising from the company's activities shall be documented in a special inventory. This could be designed with the following main points: • physical properties; • release point/source; • substance or other parameter; • quantity/concentration; • comments and assessment (possible disturbance, risk); • improvement/potential; • action/priority.
Responsibility	The line organization is responsible for carrying out investigations and assessments according to this procedure. The SHE function co-ordinates and supervises. Note: An effective assessment of SHE effects requires concentrated effort and expertise. An organization may find significant benefit in using a specialist consultant to carry out these assessments.
References	This procedure refers to the following SHE procedures: SHE POLICY SHE AUDITS SHE LEGISLATION RISK ANALYSIS SHE PERMITS

Chemicals Handling Inc., SHE Management Procedures **Procedure for SHE Reporting**				
Reg. No: SHE-4.2		Revision: 1	Valid from:	Page 1 of 2
Approved by:		Date:		Issued by:
Distribution:				

Objective	The objective of this procedure is to define the scope of and to regulate the recurrent and comprehensive reporting of SHE issues for Chemicals Handling Inc.
Scope	The following general SHE reporting is regulated in this procedure: *Externally:* • official reports to the environmental authorities (or equivalent); • information to the public/customers. Reporting to the authorities in connection with abnormal events (accidents and incidents) shall be done in accordance with the procedure for ACCIDENTS, INCIDENTS AND DISTURBANCES, SHE-11.2. *Internally:* • SHE report, quarterly summary; • SHE report, annual summary.
Principles and methods	The reporting of SHE issues shall in principle be done within the framework of each separate SHE procedure. Some general issues are reported and assembled both externally and internally (see 'Scope'). **External reporting** The official environmental report to the county government shall follow the standard prescribed in regulation XX:YY. **Annual report/environmental report** There shall be an annual summary report which, in addition to the contents of a quarterly report (defined below), should contain an analysis of conditions such as trends, targets and fulfillment of targets, causes of deviation. The annual report shall be designed so that it can be distributed both internally and externally as a SHE balance report for Chemicals Handling Inc. For requirements according to EMAS, see EMAS article 5. **Internal reporting** An internal SHE information/report to all employees shall be issued once per quarter. It shall contain among other things: • performance vs targets; • emission figures (air, water, wastes, etc.); • accidental environmental releases; • accidents and 'near-misses' (lost-time incidents, etc.); • accomplished projects, special measurements, SHE rounds, etc.; • consumption of raw material, energy, water, etc.
Responsibility	The site manager/managing director has overall responsibility for ensuring that the reporting to the authorities is carried out according to this procedure. The SHE function prepares and summarizes the material. The responsibility for the internal report lies with the SHE function.

Chemicals Handling Inc., SHE Management Procedures

Procedure for SHE Reporting

Reg. No: SHE-4.2		Revision: 1	Valid from:	Page 2 of 2
References	This procedure refers to the following SHE procedures:			

SHE POLICY   TRAINING
SHE OBJECTIVES AND ACTION PLANS SHE ROUNDS
MANAGEMENT REVIEW   OCCUPATIONAL HEALTH
SHE LEGISLATION   ENVIRONMENTAL CONTROL
SHE PERMITS   ACCIDENTS, INCIDENTS AND DISTURBANCES

Chemicals Handling Inc., SHE Management Procedures **Procedure for Health Care**				
Reg. No: SHE-5.1	Revision: 1	Valid from:		Page 1 of 2
Approved by:	Date:		Issued by:	
Distribution:				

Objective	The objective of this SHE procedure is to define the general principles and goals for health care within Chemicals Handling Inc., and the requirements of the medical resources.
Scope	This procedure applies to all activities and all employees of Chemicals Handling Inc.
Principles and methods	Chemicals Handling Inc.'s objective is that all employees shall be in good health in all aspects and in no way become negatively influenced by the working conditions at the company. A general objective is to carry out preventive health care.

In order to fulfil this ambition, a regular employee health check and a control of workplace conditions are needed. The work shall be governed by the following two main principles:

- where the assessment of SHE effects shows that the materials and the way they are handled could lead to ill-health effects, expert advice should be obtained (a risk assessment will give rise to a risk inventory);
- where necessary, regular medical checks shall be instituted.

Resources

Health care centre
Chemicals Handling Inc. utilizes the health care resources available locally according to conditions in a contract. There shall be opportunities for making basic health checks when hiring new employees.

Miscellaneous
In order to reinforce resources in the case of an acute emergency, Chemicals Handling Inc. shall have some employees trained as first-aiders. Every department shall have a first-aider who receives regular retraining. There shall be a first aid cabinet/box in all departments, and facilities for resting.

Workplace surveys
Regular surveys of the workplace shall be carried out (in accordance with the SHE procedure for OCCUPATIONAL HEALTH, SHE-9.4). This shall include, among other things, measurements of chemicals in the air, noise, illumination conditions and other ergonomic conditions based on the risk inventory. The SHE function of the company is responsible for ensuring that these checks are made.

Extent of health checks

New employees
All persons to be employed by the company shall undergo a health check before confirmation of employment.

Regular health checks
Regular health checks shall be performed. The frequency of these checks shall be agreed between the company and the health care centre.

The company shall have resources or be in a position to acquire resources when the need to perform specific detailed investigations arises. Such specific investigations should be based on the survey which should be made regularly according to the SHE procedure for OCCUPATIONAL HEALTH, SHE-9.4.

Chemicals Handling Inc., SHE Management Procedures **Procedure for Health Care**			
Reg. No: SHE-5.1	Revision: 1	Valid from:	Page 2 of 2

Principles and methods (cont'd)	**Psycho-social conditions** The company considers the psycho-social conditions of its activities to be very important. This shall be integrated into daily work and training. **Rehabilitation** Employees who, due to illness or work injury, have impaired work ability, shall be given the opportunity to return to their job or to other suitable work via rehabilitation. The relevant line manager and the personnel department handles these cases and their actions are governed by the company rehabilitation policy. **Drug tests** The company can introduce drug testing of personnel working at the company site. In addition the company will make available appropriate support for employees including support for drug and alcohol problems. **Health/fitness programme** Chemicals Handling Inc. shall promote the health and fitness of its employees through a health care centre, health/fitness programmes and regular campaigns on food habits, physical exercise, smoking and alcohol. **Documentation and follow-up** In addition to well documented files for each employee, the health care centre shall produce an annual report for Chemicals Handling Inc. The SHE function will produce summaries and analyses of work-related injuries. Absenteeism due to illness and personnel turnover shall also be documented and followed up by the personnel function. Registers of exposure to certain substances may be needed. The health care centre shall perform or let perform epidemiological investigations related to the company handling of certain chemicals when necessary. **Secrecy** The rights of the individual to keep their health files confidential will be observed.
Responsibility	The overall responsibility for Chemicals Handling Inc.'s health care lies with the site manager. The personnel manager is responsible for liaison with the health care centre. Each line manager is responsible for routine work, such as initiating workplace surveys.
References	This procedure refers to the following SHE procedures: SHE POLICY ORGANIZATION SHE OBJECTIVES AND ACTION PLANS MANAGEMENT REVIEW SHE LEGISLATION SHE EFFECTS/IMPACTS SHE REPORTING TRAINING HANDLING OF CHEMICAL PRODUCTS MANUFACTURING METHODS/R&D WORK/SCALE-UP SHE AUDITS RISK ANALYSIS SHE ROUNDS OCCUPATIONAL HEALTH ACCIDENTS, INCIDENTS AND DISTURBANCES PRODUCT CONTROL

Chemicals Handling Inc., SHE Management Procedures **Procedure for Training**				
Reg. No: SHE-5.2		Revision: 1	Valid from:	Page 1 of 4
Approved by:		Date:		Issued by:
Distribution:				

Objective	The objective of this procedure is to define the framework for competence and training from a SHE point of view for the personnel within Chemicals Handling Inc.
Scope	This procedure applies to all units within Chemicals Handling Inc. The procedure shall cover all training requirements connected with SHE. The educational levels required for various categories of personnel are defined.
Definitions	Management personnel means personnel are above the supervisor level. Training is normally the responsibility of the line manager.
Principles and methods	Well educated and trained personnel, conscious of their responsibilities, are the most important factor in maintaining a high level of competence within the system. To fulfil this the following are required: • careful selection when recruiting personnel; • a good induction system for new employees; • basic training; • regular and continual training; • follow-up to ensure that training is achieving the desired result. **Basic requirements for recruiting** When recruiting new personnel certain basic requirements for employment shall be fulfilled. In addition, opportunities for further development in the company can sometimes be determined at this point. The organization's commitment to SHE should be discussed during the employment interview, and the attitude of the applicant should be assessed. 　The requirements for each position are defined in job descriptions and associated SHE aspects have been defined for each activity and position. **Legislatively compulsory training** Some work requires evidence of certain training (work with specific substances or machines). The employer shall give this specific training in accordance with the legislation. Such training is included as a *minimum level* of the Chemicals Handling Inc. training programme. **Training programme** Personnel should receive effective and thorough training in their respective positions if they are to behave in a responsible way in terms of SHE matters. 　This procedure includes general training requirements as well as pure SHE issues. All training activities should be formulated with objectives, i.e. after final training the participants should have reached a pre-set competence level. 　The following shall be included in the overall training programme: • induction training; • training of management personnel; • supervisor training; • job-specific training for all other employees;

Chemicals Handling Inc., SHE Management Procedures **Procedure for Training**				
Reg. No: SHE-5.2	Revision: 1	Valid from:		Page 2 of 4

Principles and methods (cont'd)

- recurrent training of all personnel (especially SHE training);
- training of personnel with specific SHE functions;
- training of external personnel/contractors/transporters;
- training of customers.

Induction training

All employees shall have induction training, including:

- safety, health and environmental information on the first day of employment;
- information about the company and its products, and about general SHE activities;
- an induction programme specific to each department.

The company has developed a check-list for managers giving the induction training.

Training of management personnel

Further training of management personnel is primarily governed by individual needs and objectives. Programmes for this shall be discussed at the yearly planning/development talks. There should be specific emphasis on safety, health and environmental competence.

Supervisor training

Besides requiring technical competence in all areas of responsibilities, supervisors also need training in areas such as personnel management, and responsibilities. Within the SHE area, occupational health training shall be conducted.

Job-specific training for all other employees

The training of employees shall be performed within each unit or department. There should be a structured programme with all the various training elements. The employee shall have a clear understanding of the processes and work courses in order to act in a knowledgeable and competent way. Therefore the programme shall contain substantial elements of a theoretical and explanatory nature in addition to the practical elements. There shall be a special identified tutor for this training.

A follow-up/competency test shall be conducted before an employee is permitted to work independently.

Basic specific SHE training

All personnel shall receive one day's basic training in SHE issues.
This shall include:

- environmental issues;
- legislation for environmental, occupational health and chemicals matters;
- Chemicals Handling Inc. and SHE (SHE policy, SHE management system, permits, conditions);
- openness and activities to create confidence;
- specific SHE issues for Chemicals Handling Inc.

General recurring SHE training

All employees of Chemicals Handling Inc. shall have SHE training corresponding to at least half a day per year. Some of this training shall be exercises in evacuation, fire-fighting and use of personal protection equipment. The training shall also contain elements which shall be decided year by year by the management team. Examples of training in this category are courses in first aid, environmental issues, ecology, better working conditions, chemical health risks, etc. Production personnel shall have additional training of at least half a day yearly, to include larger exercises.

Chemicals Handling Inc., SHE Management Procedures **Procedure for Training**				
Reg. No: SHE-5.2	Revision: 1	Valid from:		Page 3 of 4

| **Principles and methods (cont'd)** | *Training for special SHE personnel*
The company ensures that all employees with special tasks within the SHE area receive the necessary training to perform their tasks effectively. One such category of employees is safety representatives.

SHE training of contract employees
All external workers within the Chemicals Handling Inc. premises or other areas of responsibility, must have received basic induction safety and environmental training in workplace risks and the emergency plan. This training shall be repeated once per year. Each manager of an external workforce is responsible for initiating the training. The actual training is performed mainly by the SHE function. Completed training shall be documented and confirmed and acknowledged by the external person.

Customer information/training
It is the responsibility of Chemicals Handling Inc. to ensure that customers are adequately informed/trained to be able to handle the company's products in an acceptable way from a SHE point of view. The company's products shall always have clear SHE information attached to or otherwise associated with them. In some cases the customers shall be invited to Chemicals Handling Inc. for information/training and in other cases the company shall train the customer at their premises. The SHE procedure for PRODUCT CONTROL, SHE-12.1, regulates this question in more detail.

Resources
Training shall mainly be performed internally with internal and in some cases external resources. Resources for planning and carrying out the training shall be allocated in the departments.
For training workers in the practical aspects of their job, there shall be dedicated instructors with a high level of competence.

Training material
All essential training material should be in the written form. The person in charge of employee development/ training has overall responsibility for the training material for the company common parts, particularly for introduction training.
The responsibility for departmental training material lies with the appropriate person within the department. This responsibility includes keeping the training material up-to-date.

Planning
Besides general planning for the systematic training of categories of employees, training needs shall also be identified and planned on an individual basis at development/planning talks held once a year between the employee and their immediate superior.
Where an employee has changed work tasks, the necessary training shall be planned and scheduled in advance and checked as necessary. This is part of management of change.

Competence assessment
All training shall be followed up by an assessment of the results and reception. Employees in key positions shall be subject to a formal competence check after training and before being allowed to work independently.
The follow-up ensures that each individual has acquired an adequate competence level and that the training has been of adequate extent and quality. The competence assessment shall be made by the person responsible for the respective training. |

Chemicals Handling Inc., SHE Management Procedures **Procedure for Training**				
Reg. No: SHE-5.2	Revision: 1	Valid from:		Page 4 of 4

Principles and methods (cont'd)	**Documentation** Training programmes shall be documented. It should be clear from the individual employee's documentation which training has been performed and which competencies have been met. This documentation shall be treated, updated and filed by the line manager responsible for personnel.
Responsibility	Overall responsibility for the extent of training, training programmes, quality levels, etc. lies with the site manager. Department managers and equivalent persons are responsible for carrying out the training according to the programme and for controlling and maintaining an adequate competence level within their employees.
References	This procedure refers to the following SHE procedures: SHE POLICY NON-PROCESS HAZARDS SHE LEGISLATION INVESTMENT PROJECTS INTERNAL COMMUNICATION MANAGEMENT OF CHANGE EXTERNAL COMMUNICATION MANUFACTURING METHODS/R&D WORK/SCALE-UP PURCHASING TECHNOLOGY SALE GENERAL SITE RULES EMERGENCY RESPONSE INSTRUCTIONS ENVIRONMENTAL CONTROL WORK PERMITS ACCIDENTS, INCIDENTS AND DISTURBANCES CONTRACTORS FIRE PROTECTION HANDLING OF CHEMICAL PRODUCTS TRANSPORT HANDLING OF WASTES PRODUCT CONTROL SOIL AND GROUNDWATER PROTECTION

Chemicals Handling Inc., SHE Management Procedures **Procedure for Internal Communication**				
Reg. No: SHE-5.3		Revision: 1	Valid from:	Page 1 of 1
Approved by:		Date:		Issued by:
Distribution:				

Objective	The objective of this procedure is to communicate the most important principles and methods of the SHE policy of Chemicals Handling Inc. throughout the organization to promote commitment and motivation.
Scope	This procedure applies to all activities within Chemicals Handling Inc. It is aimed at all employees, but the main target group is managers.
Principles and methods	The following principles shall be valid: • communication regarding safety, health and environment shall be correct, unambiguous and based on factual information; • it should be ensured that the information has been correctly received; • for important issues the information supplier shall ask for formal confirmation of the transferred information and, in particularly important cases, in written form; • the aim is to inform in such a way that the receiver fully understands the relationships and contexts – *why* and not only *that* or *how* a task shall be performed; • SHE issues shall always be on the agenda for board meetings, management group meetings, departmental meetings, workplace meetings and SHE committee meetings; • managers shall be sensitive to employees' views on SHE issues; • management involvement shall be visible to the employees; • management shall set good examples relating to SHE; • it is part of every manager's duty (particularly within the SHE area) to act with 'Responsible Care' towards the employees and the environment; • management must intervene when existing rules on SHE behaviour are violated; • management shall actively follow up accidents, incidents and disturbances and ensure that the event can not be repeated; • SHE training shall be performed both internally and externally and by employee participation in external courses; • information on company performance within the SHE area shall be easily accessible by all employees and shall be transmitted by the managers in a suitable forum; • Chemicals Handling Inc. shall have a system to motivate employees to come up with ideas for improvements within the SHE area; • the company shall inform and motivate personnel to achieve higher targets; • exchange visits will be arranged with other companies and organizations within the SHE area. SHE issues shall be included in the one-to-one development talks held by managers with employees. The type of information that should be given can be seen in the procedure for SHE REPORTING, SHE-4.2. A record of any talks conducted should be kept.
Responsibility	Every manager is responsible for communicating information according to the above guidelines. Every employee is responsible for participating constructively in the meetings and the talks as defined above.
References	This procedure refers to the following SHE procedures: SHE POLICY SHE REPORTING SHE LEGISLATION TRAINING

Chemicals Handling Inc., SHE Management Procedures **Procedure for External Communication**				
Reg. No: SHE-5.4		Revision: 1	Valid from:	Page 1 of 2
Approved by:		Date:		Issued by:
Distribution:				

Objective	The objective of this procedure is to communicate the most important principles and methods with the aim of achieving good relationships with external stakeholders and creating a picture of Chemicals Handling Inc. as a safe and environmentally-conscious company.
Scope	This procedure concerns all types of external information on SHE issues passed between Chemicals Handling Inc. and external stakeholders.
Principles and methods	The following principles should be valid: • openness shall govern as far as company secrecy (in R&D, manufacturing and business) allows; • the spirit of the SHE policy and commitment to Responsible Care shall be fulfilled; • Chemicals Handling Inc. shall be a good neighbour; • communication from Chemicals Handling Inc. shall always be active; • employees shall be the 'ambassadors' of the company. Chemicals Handling Inc. has a number of external parties with which information is regularly exchanged: *Authorities* Authorities shall be kept informed about the ongoing activities according to agreed programmes and reporting. They shall be given preliminary information about changes or new projects at an early stage, and shall be kept informed. The objective shall be to give a clear overview of the planned conditions. *Community bodies* Community bodies shall essentially have the same information as the authorities. Once every second year, or more frequently if necessary, the community shall be invited to the company and informed about its activities. *Schools* Chemicals Handling Inc. shall strive to inform schools and other educational institutions about the company's work in the SHE area. *Local media* The local media shall always be given information speedily and based on facts. The relationship with the local media shall be based on personal relationships. *Neighbours* The company's neighbours shall regularly be given general information about the conditions and possible changes within Chemicals Handling Inc. The company's SHE policy shall be available to the public. Use will be made of leaflets, 'open days', schools liaison, etc. *Environmental groups* Environmental groups shall be given relevant, factual information about conditions at the company. The company shall aim to co-operate with such groups and will consider inviting representatives from these groups to the company for discussions and tours. *Customers* Customers shall be given relevant and full information and training about the company products and its way of handling SHE matters as required.

Chemicals Handling Inc., SHE Management Procedures **Procedure for External Communication**				
Reg. No: SHE-5.4		Revision: 1	Valid from:	Page 2 of 2

Principles and methods (cont'd)	*Own employees*
	General communication within Chemicals Handling Inc. is dealt with in the procedure for INTERNAL COMMUNICATION, SHE-5.3. Employees shall be given sufficient training and information about SHE issues to be able to act as 'SHE ambassadors' for the company. This means that they shall, among other things, be informed about how actual emissions from the company and accident frequencies compare with permissible values or targets.
	    Union representatives should be given the chance to be present at visits by authority representatives.
	
	Official contact persons
	Official contacts with external parties must be able to give factual, unambiguous and, in some cases, continuous information according to the principles above. Therefore, they shall always be handled by especially appointed and trained personnel as defined in the following:
	
	• community emergency services and national authorities for explosives, flammable materials and for rescue services: SHE function;
	• community health and environmental protection authorities: SHE function;
	• regional and national environmental protection authorities: SHE function;
	• regional and national authorities for occupational health: SHE function;
	• media: site manager.
	
	The information shall, in relevant cases, be agreed with the site manager.
	
	Handling of complaints/disturbances/inquiries (excluding company products)
	Contact with external groups, in these cases, shall be handled by a person appointed by the company site manager/managing director. Complaints from people in the surrounding area about Chemicals Handling Inc.'s activities shall always be taken seriously and handled speedily. Disturbances shall be handled in the same way. Every complaint/disturbance shall be documented and filed.
	    The issues surrounding the handling of information in an emergency case shall follow the same principles. This is covered in Chemical Handling Inc.'s emergency plan.

Responsibility	The site manager has overall responsibility for adhering to the contents of this procedure. Besides this the responsibility is defined above.
References	This procedure refers to the following SHE procedures: SHE POLICY TRAINING SHE LEGISLATION INTERNAL COMMUNICATION SHE PERMITS

Chemicals Handling Inc., SHE Management Procedures **Procedure for Purchasing**				
Reg. No: SHE-6.1	Revision: 1	Valid from:		Page 1 of 3
Approved by:	Date:		Issued by:	
Distribution:				

Objective	This procedure shall ensure that all purchased materials meet SHE requirements, whilst maintaining the high performance of the purchased goods for the company activities.
Scope	This procedure applies to the whole of Chemicals Handling Inc.'s purchasing activities where SHE effects can be possible, such as the purchase of: • chemical products; • equipment; • other goods; • services.
Principles and methods	The SHE issues concerning all potential purchases shall be checked beforehand by a competent party, possible alternatives shall be considered (substitution principle), and SHE aspects shall be compared with economic aspects. Only goods which fulfil formal SHE requirements shall normally be purchased. This means, for example, that chemicals on 'exception lists' and non-CE marked equipment shall be avoided. SHE information shall always be given to the relevant personnel involved, before the purchased product, equipment or service is used. Purchases shall only be made from suppliers who have previously shown that they work seriously with SHE issues, or can be shown to be serious at a review, and have been approved according to a special procedure. When purchasing new services or goods the supplier shall be reviewed to ensure that it can meet company SHE requirements before being added to the list of suppliers. **Work procedure** A simple work procedure, which is common to all types of purchases, is outlined below. The work procedure is shown in Attachment 1. The particular requirements and handling which could be requested for different types of purchases (chemical product, equipment, service, etc.) are specified under each point. *1) Purchase request* The person placing the order, normally the department manager, the supervisor or an appointed project leader, submits a written purchase request and a technical specification to the purchasing function. The person placing the order for a new product or service is responsible for ensuring that all SHE aspects of the product or service have been evaluated. *2) Review* A new chemical product shall be reviewed by the SHE function or, in some cases, the SHE committee before an external order is placed. Products which have already been handled do not need to be reviewed, unless they are to be used in a new way. The SHE hazards of all new equipment shall be assessed. A formal risk analysis may be needed according to the procedure for RISK ANALYSIS, SHE-9.2, when introducing a new chemical product or new equipment. For equipment, CE labelling can be required. *3) Order* The purchasing function places the order and requests all relevant SHE information from the supplier. Product information must be available at this point and certain information shall be delivered before the goods. Functional guarantees and liability insurance shall be defined.

Chemicals Handling Inc., SHE Management Procedures **Procedure for Purchasing**				
Reg. No: SHE-6.1		Revision: 1	Valid from:	Page 2 of 3

Principles and methods (cont'd)	*4) Preparation of SHE instructions and information* The purchasing function sends the supplier information to the person who placed the order. When purchasing goods which are already in use and have the relevant SHE information, the person placing the order need only check that this is still relevant. When purchasing new goods the person placing the order shall develop the necessary information and training material (such as operating instructions and safety instructions). Supplier material shall always be adjusted to fit the actual situation and application by this person before it is used for training. The SHE assessment may need updating. *5) Training and/or control of training* Before the goods are used and normally before delivery, the person who placed the order, or the relevant line manager, shall ensure that all personnel that could be affected by the introduction or use of the goods will receive suitable and adequate training based on the instructions and information prepared at the previous stage. *6) Delivery control* When the goods arrive at Chemicals Handling Inc. they should be checked according to an approved procedure to ensure that the specified requirements are fulfilled and that all relevant SHE information is available. If this is not the case, it should be requested from the supplier. *7) Use/application* Provided that the above conditions are fulfilled, the goods can be used with due consideration to the company's SHE procedures.
Responsibility	Overall responsibility to maintain adherence to this procedure lies with the purchasing manager. Besides that, responsibility is specified above and in Attachment 1.
References	This procedure refers to the following SHE procedures: SHE POLICY CONTRACTORS SHE LEGISLATION MANAGEMENT OF CHANGE TRAINING INVESTMENT PROJECTS INSTRUCTIONS RISK ANALYSIS PLANT INTEGRITY AND MAINTENANCE

Chemicals Handling Inc., SHE Management Procedures
Procedure for Purchasing
Attachment 1 — Work procedure for purchasing

Reg. No: SHE-6.1	Revision: 1	Valid from:	Page 3 of 3
			Attachment 1

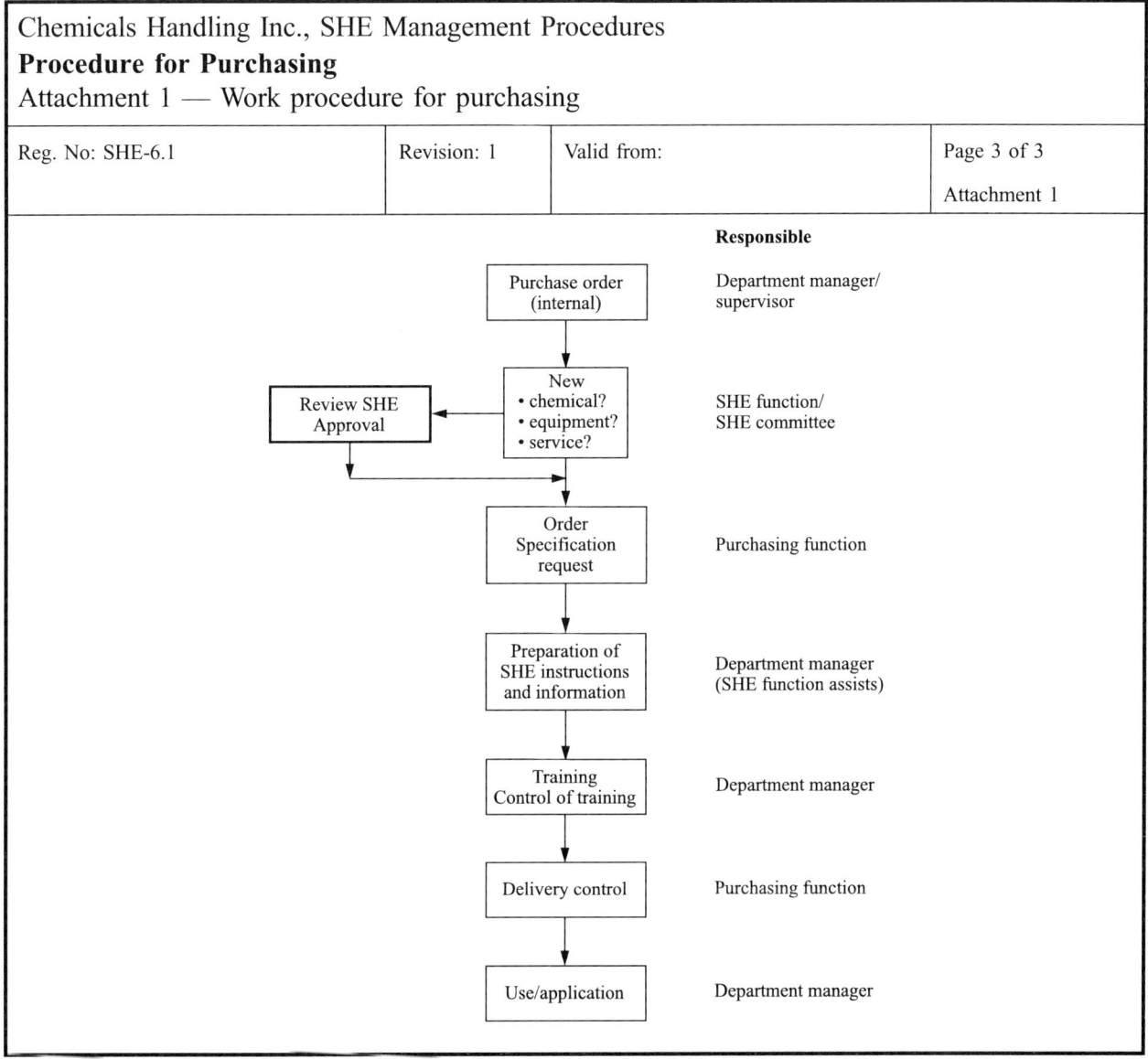

Responsible

Purchase order (internal) — Department manager/ supervisor

New
• chemical?
• equipment?
• service? — SHE function/ SHE committee

Review SHE Approval

Order Specification request — Purchasing function

Preparation of SHE instructions and information — Department manager (SHE function assists)

Training Control of training — Department manager

Delivery control — Purchasing function

Use/application — Department manager

Chemicals Handling Inc., SHE Management Procedures **Procedure for General Site Rules**				
Reg. No: SHE-7.1	Revision: 1	Valid from:		Page 1 of 2
Approved by:	Date:		Issued by:	
Distribution:				

Objective	The purpose of this procedure is to define the general site rules, which form the basis for the successful running of Chemicals Handling Inc.'s activities from a SHE perspective.
Scope	This procedure applies to all activities within the Chemicals Handling Inc. industrial site and is also partly applicable to office work and to transportation of products. All personnel present in the industrial area – the company's personnel, temporary external personnel and visitors – are covered.
Principles and methods	General order is the single most important condition for a safe and environmentally-friendly way of working. The following rules for the industrial site are compiled in a folder which should be handed over to all external personnel who will work within this area. **Conditions of entry to and exit from the site** *Personnel* • Only authorized personnel may enter the industrial area and remain there. • Authorized personnel include the company's employees, and other personnel and visitors who need to be in the industrial area for their work. • Non-Chemicals Handling Inc. personnel shall normally be accompanied by a responsible person from the company. Department managers can make exemptions from this for personnel who are very familiar with the conditions and rules within the industrial area. • Juveniles may not enter the industrial area without special permission from the site manager. • All personnel who pass in or out of the gates of the industrial area shall be registered by electronic or other means. *Vehicles* • Vehicle traffic within the industrial area shall be kept to a minimum. Vehicles shall have special permits for entrance and thereby be registered. The SHE function may issue certain vehicles which are frequently required within the area, with a permanent vehicle pass. • A special driving permit (hot work permit) is needed for entrance into defined parts of the industrial area. • The maximum speed allowed within the industrial area is 30 km/h. **Remaining on the industrial site** *General* • Personnel shall not unnecessarily stay in areas with increased risks, such as production. • Smoking is prohibited within the industrial area except where a sign saying 'Smoking allowed' is exhibited. • Personal protection equipment (e.g., safety helmets, safety glasses, ear protection, safety shoes) must be worn in defined areas. • People under the influence of drugs are not allowed into the industrial area. Alcoholic beverages or other drugs may not be brought onto the industrial site. • Photography and filming is prohibited without special permission from the site manager. • In the case of an emergency situation, the emergency plan comes into effect and all work permits are cancelled. • Any person who does not follow the issued rules can be evicted from the industrial area.

Chemicals Handling Inc., SHE Management Procedures **Procedure for General Site Rules**				
Reg. No: SHE-7.1		Revision: 1	Valid from:	Page 2 of 2

Principles and methods (cont'd)	*Work* • All work on equipment, except normal operating work, needs a work permit according to a special procedure. • Everyone shall keep their workplace clean and free from chemical spills, tools, scrap, etc., in order to facilitate SHE work and for general comfort. • Work is finished only when clean-up has been completed. • A special procedure for disposal of wastes shall be followed. *Study visits* • Study visits shall be approved by the production manager. • External person(s) must always be accompanied by a person from Chemicals Handling Inc. • Visits shall be performed in such a way to minimize disturbance to normal work.
Responsibility	It is everybody's responsibility to keep the workplace tidy and to follow the above rules. The formal responsibility for adhering to these rules lies with every supervisor and every department manager.
References	This procedure refers to the following SHE procedures: SHE POLICY HANDLING OF CHEMICAL PRODUCTS TRAINING NON-PROCESS HAZARDS WORK PERMITS EMERGENCY RESPONSE CONTRACTORS

Chemicals Handling Inc., SHE Management Procedures **Procedure for Instructions**				
Reg. No: SHE-7.2	Revision: 1	Valid from:		Page 1 of 2
Approved by:	Date:		Issued by:	
Distribution:				

Objective	The objective of this procedure is to define the necessary requirements on Chemicals Handling Inc.'s work instructions in order to perform sufficiently from a SHE point of view.
Scope	This procedure applies to all activities within Chemicals Handling Inc.
Principles and methods	All work within Chemicals Handling Inc. shall in principle have some form of instructions. They need not always be written; good training will in many instances replace the need for detailed written instructions. Instructions shall be available for work that can affect safety, health and environment, give rise to disturbances and so on. This applies both to routine work and to unique or infrequent work, but which could result in risk situations, such as: • production (all phases including start-up, normal operation, shutdown and emergency situations); • maintenance; • laboratory work; • other handling or transport of chemicals. In order to ensure that work activities – under normal circumstances or in extraordinary situations – are carried out safely there should be *one* predetermined and approved way for the activity to take place. Instructions are needed as documentation and as a basis for training in the correct and safe performance of a work activity. **Design of instructions** Instructions should normally be in writing, but verbal instructions will suffice for less risky operations. SHE aspects shall normally be integrated in the work instructions with a clear statement of the necessary safety precautions and measures. In certain cases a separate SHE instruction may be justified. Instructions should be written in a short form, preferably as check-lists. Every instruction shall contain: • objective; • scope; • method – detailed contents; • hazards and means of control. The instructions shall clearly define such things as permissible values of operating parameters. In particular, those criteria which will lead to special measures or reporting with respect to SHE shall be stated. Instructions should be developed in co-operation with the personnel who are to use them, and be written in a style suitable for the user. **Updating and revision** Instructions shall be revised on a regular basis (once per year) and be updated as necessary. **Document control** Every instruction shall be identifiable and be part of a document control system. **Information and training** Instructions shall be easily accessible by users. There should be a procedure in the system to ensure that all relevant employees (e.g., an operator in a particular position on a shift) are aware of the information contained in a new or revised instruction. Instructions should be included as an aid in the scheduled training of personnel (see procedure for TRAINING, SHE-5.2).

Chemicals Handling Inc., SHE Management Procedures			
Procedure for Instructions			
Reg. No: SHE-7.2	Revision: 1	Valid from:	Page 2 of 2
Responsibility	The respective department manager has overall responsibility for the instructions and adherence to them.		
	Supervisors are responsible for ensuring that, within their respective areas of responsibility, instructions exist for all work which requires them. Supervisors are also responsible for ensuring that all employees receive training in the instructions.		
	Each employee is responsible for following the instructions. If this is not possible, the next higher supervisor or manager shall be contacted for their decision.		
	There should be an observation programme to ensure that workers follow instructions.		
References	This procedure refers to the following SHE procedures:		
	SHE POLICY WORK PERMITS		
	SHE LEGISLATION CONTRACTORS		
	TRAINING HANDLING OF CHEMICAL PRODUCTS		
	INTERNAL COMMUNICATION NON-PROCESS HAZARDS		
	PURCHASING MANAGEMENT OF CHANGE		
	GENERAL SITE RULES INVESTMENT PROJECTS		
	PLANT INTEGRITY AND MAINTENANCE MANUFACTURING METHODS/R&D WORK/SCALE-UP		

Chemicals Handling Inc., SHE Management Procedures **Procedure for Plant Integrity and Maintenance**				
Reg. No: SHE-7.3	Revision: 1	Valid from:		Page 1 of 5
Approved by:	Date:		Issued by:	
Distribution:				

Objective	The objective of this procedure is to ensure that through effective inspection and maintenance systems, the integrity of all equipment and facilities is maintained.
Scope	The procedure covers the whole of Chemicals Handling Inc.'s activities and applies to maintenance of certain equipment which could give SHE effects if it were to fail. It also applies to the control and maintenance of those units or equipment which are specifically intended for the protection of safety, health and environment. The company maintenance system contains many parts and is not an integral part of the SHE management system. Only those parts which are referred to in this procedure are considered to be part of the SHE management system. This includes, among other things, the legislative requirements on official controls and the company's own controls concerning SHE.
Principles and methods	A basic prerequisite for a safe plant is a good programme for mechanical and other functional integrity of the equipment. The basis is laid for this during design, engineering and construction of the plant. The company has a set of standards for the design, engineering and construction of its facilities. This is covered in the procedure for ENGINEERING STANDARDS, SHE-8.1. There are legislative requirements for mechanical and electrical equipment in most countries and these shall be the basic requirements adopted by Chemicals Handling Inc. The equipment shall also fulfil the EU directives and other EU legislation when relevant. Maintenance is a vital activity and should therefore contribute towards fulfilling the overall company objectives. Maintenance is also a very important factor for safety, health and environment. Deficiencies in maintenance have often shown to be a main cause of failure with consequences for SHE. Technical systems which are primarily there for safety, health or environmental purposes shall be maintained so that they are available for fulfilling high SHE performance requirements, i.e. often higher than the production system itself. For the production systems to work well, operations personnel must know the maintenance requirements. Maintenance personnel should also be aware of the operational requirements of the equipment. This mutual understanding and knowledge of the requirements shall be provided by training. Chemicals Handling Inc. shall give priority to preventive and predictive maintenance to bring down acute repair maintenance to a minimum. The Managing Director shall ensure that appropriately qualified professional or technical staff are appointed to assume overall responsibility for the following functions: • pressurized systems; • lifting equipment; • structures and civil work; • electrical equipment; • control systems and process-related computers. Appropriately qualified staff may be appointed to fulfil more than one of the above functions. A formal record of the appointments and areas of responsibility shall be made. The appointed staff may be either Chemicals Handling Inc. employees or, where a full time appointment is not appropriate, a consultant or contract employee. If in doubt over the relevant level of expertise required, the Managing Director shall consult the appropriate professional organization for advice.

Chemicals Handling Inc., SHE Management Procedures **Procedure for Plant Integrity and Maintenance**			
Reg. No: SHE-7.3	Revision: 1	Valid from:	Page 2 of 5

| **Principles and methods (cont'd)** | **MAINTENANCE OF ALL TYPES OF EQUIPMENT**

Housekeeping
A basic condition for good maintenance is to have good housekeeping. This should be achieved through continuous effort and directed scheduled cleaning programmes.
Such programmes can be directed towards keeping the system functioning well and preventing failures and sudden releases.

Preventive and predictive maintenance
The appointed professional or technical staff shall ensure that programmes for testing, inspection and maintenance are developed to meet all legislative requirements. The programmes shall also cover all equipment where failure would result in a SHE incident.
  The programmes shall cover, as a minimum:

• pressure vessels;
• tanks;
• machinery;
• lifting equipment;
• instrumental protection equipment;
• programmable electronic systems;
• electrical equipment;
• structures;
• hoses and couplings;
• chemical drains.

  In preparing these programmes consideration should be given to good practice including:

• tightness tests of equipment and piping systems (including hoses, couplings, etc);
• visual controls of equipment;
• lubrication of equipment;
• vibration measurements of rotating equipment;
• thickness measurements of equipment and piping systems;
• control of safety systems (trips, high level protection, alarms, etc).

  The programmes shall clearly state the item/object, control/test activity and method, test interval and responsible person. The result of performed control/measurement and time and responsible person shall be documented and the system shall include a systematic way of gaining and feeding back experiences from maintenance work in order to be able to adjust the test interval to actual needs.
  Where relevant, requirements of authority control shall be built into the company control programme.

Identification of maintenance requirements
There shall be documented logging of operations upsets and equipment disturbances/failures (by supervisors) to identify the repair and control requirements of equipment.

Follow-up and documentation of performed maintenance activities
Safety-critical maintenance activities shall be documented. Based on these data, analyses and summaries of availability and SHE disturbances shall be made on a regular basis. |

Chemicals Handling Inc., SHE Management Procedures **Procedure for Plant Integrity and Maintenance**			
Reg. No: SHE-7.3	Revision: 1	Valid from:	Page 3 of 5

Principles and methods (cont'd)	**New installations and modifications** When installing new or modified equipment connected with maintenance work, a review of the engineering (drawings and documentation) shall be carried out before construction/erection (in accordance with the procedures for INVESTMENT PROJECTS, SHE-8.2, and MANAGEMENT OF CHANGE, SHE-7.10. In the same way checks shall also take place after construction according to the same procedures. **Performance of the work** The risk of exposure to chemicals and of unintentional releases to the environment, etc. often increases whilst performing maintenance work. A safe way of performing the work is regulated in several procedures of which the most important are: WORK PERMITS, SHE-7.4; HANDLING OF CHEMICAL PRODUCTS, SHE-7.6; and OCCUPATIONAL HEALTH, SHE-9.4. **Instructions** In principal there should be instructions for all maintenance work. In many cases general maintenance instructions (written or verbal) are adequate, but in certain cases special written instructions shall be worked out and be approved by the immediate manager, before work can commence. See also the procedure for INSTRUCTIONS, SHE-7.2. **MAINTENANCE AND CONTROL OF SAFETY EQUIPMENT** Programmes for the testing, inspection and maintenance of the following equipment shall be developed by the SHE function: • alarms; • fire-fighting equipment (fixed and mobile); • breathing apparatus; • safety equipment (e.g., safety showers); • emergency shutdown of machines; • containment systems; • abatement systems. An example of a programme for the control of safety equipment is given in Attachment 1. The controls shall be documented. Maintenance of environmental equipment shall follow the principles and methods given above for 'all types of equipment', where relevant. The control of environmental equipment is addressed in the procedure for ENVIRONMENTAL CONTROL, SHE-9.5. The controls shall be documented.

Responsibility	Overall responsibility for manufacturing (= operations and maintenance) lies with the production manager and other department managers respectively. The appointed professional staff are responsible for ensuring that all legislative requirements are met. The responsibility for identifying and performing the necessary maintenance lies with the supervisors in the production department. The responsibility for the technical performance of the maintenance work lies with the maintenance supervisor (also in the case when contractors are used).
	    The maintenance supervisor is responsible for creating the maintenance programme and documenting and analysing the maintenance activities. The work shall be carried out in close co-operation with the operations supervisors and the production manager.
	    The responsibility for following the company engineering standards lies with the project leader and then with the production manager once the plant has been taken over.

Chemicals Handling Inc., SHE Management Procedures **Procedure for Plant Integrity and Maintenance**				
Reg. No: SHE-7.3	Revision: 1	Valid from:		Page 4 of 5
References	This procedure refers to the following SHE procedures: SHE POLICY WORK PERMITS SHE LEGISLATION CONTRACTORS SHE PERMITS ENGINEERING STANDARDS TRAINING MANAGEMENT OF CHANGE PURCHASING INVESTMENT PROJECTS INSTRUCTIONS			

Chemicals Handling Inc., SHE Management Procedures
Procedure for Plant Integrity and Maintenance
Attachment 1 — An example of a programme for the control of safety equipment

| Reg. No: SHE-7.3 | | Revision: 1 | Valid from: | | Page 5 of 5 |
| | | | | | Attachment 1 |

Equipment	Control/test method	Frequency	Responsibility	Comments
Fire monitors	Operate valves Capacity test	1 per quarter 1 per year (rolling programme)	Supervisor (operations) Supervisor (operations)	All controls shall be documented
Fire extinguishers	Control at place, functional test	1 per month	Supervisor (operations) resp. administrative manager (office)	
Sprinkling equipment	Inspection	1 per year	SHE manager	
Gas alarm, fixed installation	Functional test	1 per month	Supervisor (maintenance) (instrument technician)	
Evacuation alarm	Functional test	1 per 6 months	Production manager	
Emergency showers	Functional test	1 per month	Supervisor (operations)	
Alarm, environmental releases −sewer −air cleaning	Functional test, calibration	1 per month 1 per month	Supervisor (maintenance) (instrument technician)	Calibration by accredited laboratory; 1 per year

Chemicals Handling Inc., SHE Management Procedures **Procedure for Work Permits**				
Reg. No: SHE-7.4		Revision: 1	Valid from:	Page 1 of 9
Approved by:		Date:		Issued by:
Distribution:				

Objective	The risk of releases and damage to humans and to the environment increases during repair and maintenance work. In order to be able to perform such work in a controlled and safe way, a system with formal authorization for such activities is needed. The objective of this procedure is to give the basic principles and rules within this area.
Scope	This procedure applies to all activities within Chemicals Handling Inc. All work on equipment, except normal operations work, is included in this procedure. Only the operations personnel are authorized to work without a work permit in the production units. All other personnel shall have a permit from authorized operations personnel.
Principles and methods	**Work procedure** The general work procedure is as follows: • the operations personnel prepare and clear the equipment before work and issue a permit; • maintenance or other personnel carry out the work and return the equipment with the work permit to the operations personnel; • the operations personnel check the equipment and prepare it for operational status. An issued permit states that the defined work can be carried out in a safe way, while taking into account certain conditions connected to the permit. However, it is always the responsibility of the person carrying out the work to show great care and attention during the work. **Types of work permits** *Permit to work* A permit-to-work is required for all work on site, including that of contractors, with the exception of: • operational activities; • regular maintenance activities where a procedure has been agreed and the staff trained in its application. The departmental manager shall ensure that management and supervisory staff, including relief staff, are trained to issue permits to-work. All appointments shall be in writing. Attachment 1 contains guidance on the issuing of permits-to-work. *Isolation of equipment* Where the equipment to be worked on contains hazardous materials or is connected to a source of energy, it must be effectively isolated before work starts. This applies to: • chemical hazards; • hazards of high or low pressures or temperatures; • electrical equipment; • rotating equipment. Attachment 2 contains key principles for the safe isolation of equipment. *Hot work permit* Before work which could generate heat or cause sources of ignition is carried out in areas where flammable materials could be present, a 'hot work' permit is required. (This includes work on live electrical or instrument equipment, use of 'non-flameproof' test equipment and movement of vehicles into the area.)

Chemicals Handling Inc., SHE Management Procedures **Procedure for Work Permits**			
Reg. No: SHE-7.4	Revision: 1	Valid from:	Page 2 of 9

Principles **and** **methods** **(cont'd)**	Hot work should be avoided wherever possible, e.g. by using alternative methods of carrying out the work or by moving the equipment to a workshop or other safe place. The department manager shall ensure that staff, usually one grade above those authorized to issue work permits, are trained to issue hot work permits. All appointments shall be in writing. Attachment 3 contains guidance on hot work permits. *Entry into confined spaces* No-one, including operations personnel, may enter a confined space without a confined space entry permit. No-one may enter a confined space for rescue unless wearing a suitable breathing apparatus (portable breathing apparatus set or air-fed mask). The department manager shall ensure that management staff, usually one level above those issuing permits-to-work, are trained to issue confined space entry permits. All appointments shall be in writing. Confined spaces include all vessels, tanks and receivers, drains and culverts, and pits deeper than 1.2 m. They include any area where dangerous gases, liquids or dusts could accumulate or where the atmosphere may be deficient in oxygen. Before issuing a confined space entry permit, consideration should be given to carrying out the work in a way which will not require entry. Attachment 4 contains guidance on issuing a confined space entry permit. *Excavation permit* An excavation permit is required for any excavation work on site. The site manager shall ensure that a person is trained to issue excavation permits. The appointment shall be in writing. Attachment 5 contains guidance on excavation permits. *Other hazards* In some cases additional permits procedures may be required to control specific hazards on a site – for instance, work with radioactive materials, or work on high voltage electrical equipment. **Common regulations** A work permit is normally issued to a maintenance supervisor or technician employed by Chemicals Handling Inc. When the work is carried out by external personnel there should always be a person from Chemicals Handling Inc. responsible for the work. The permit is issued to the responsible person within Chemicals Handling Inc., who is responsible for checking that the contractor knows the general safety rules, and for informing the contractor about the special rules and precautions for the work. **Performance** *Preparations* Safety during maintenance work depends on how well the work has been prepared, and the operator for the respective area is mainly responsible for this. The work area shall be clean from combustible materials and irrelevant objects. Electrical motors for rotating equipment shall be stopped and locked according to a special instruction. The equipment shall normally be free from combustible materials, gases and other chemicals and blinded or completely isolated from the other systems in operation. *Detailed execution* It is important to observe great caution during the execution of the work because equipment can be pressurized or contain chemicals despite careful preparations. Wearing of personal protection equipment is regulated in the procedure for HANDLING OF CHEMICAL PRODUCTS, SHE-7.6, and should be specified for each task on the permit form. If the work cannot be considered as completely without risk, continuous guarding shall be carried out (also during pauses) by, for example, gas testing and fire guards.

Chemicals Handling Inc., SHE Management Procedures **Procedure for Work Permits**			
Reg. No: SHE-7.4	Revision: 1	Valid from:	Page 3 of 9

Principles and methods (cont'd)	*After-control* The workplace shall be returned to its original state when work has been completed. After completion of hot work, the workplace shall normally be guarded for $1-2$ hours. Thereafter the unit shall be restored to operational mode by or under the supervision of the operations personnel (removing of blinding, tightness test, flushing, etc.). Attachment 6 gives some general guidelines on preparations before and after maintenance work. **Training** All personnel, both internal and external, who are concerned with work permits of any type, shall have relevant training. This shall be part of the basic training for operations and maintenance personnel. Personnel who have the right to issue permits shall have special training, including on responsibility aspects. Personnel who carry out hot work shall be specially trained for this.
Responsibility	The overall responsibility for this procedure lies with the production manager. Besides this the responsibilities are defined above.
References	This procedure refers to the following SHE procedures: SHE POLICY GENERAL SITE RULES SHE LEGISLATION PLANT INTEGRITY AND MAINTENANCE TRAINING CONTRACTORS

Chemicals Handling Inc., SHE Management Procedures
Procedure for Work Permits
Attachment 1 — Guidance on issuing permits-to-work

Reg. No: SHE-7.4	Revision: 1	Valid from:	Page 4 of 9
			Attachment 1

When issuing a permit-to-work all the following aspects shall be considered and specified:

- is the work to be done clearly defined?
- is the equipment to be worked on clearly defined?
- what are the hazards and risks?
- how will these be controlled?
- is the equipment properly isolated? (see Attachment 2)
- will special precautions be required?
- what safety equipment is required, including personal protective equipment?

Permits-to-work must be accepted by the person doing the work or their supervisor. The period of validity must be specified on the permit. The permit is valid between the time limits stated on the form, with a maximum of eight hours. If the work is not started within two hours from the stated starting time or is interrupted for more than two hours, the permit becomes invalid. In case of a gas or fire alarm, the permit becomes invalid and the work shall be interrupted immediately.

  Copies of all the permits in force must be kept in the control room. The permit must be signed off and returned to the control room on completion of the work.

Chemicals Handling Inc., SHE Management Procedures **Procedure for Work Permits** Attachment 2 — Key principles for safe isolation			
Reg. No: SHE-7.4	Revision: 1	Valid from:	Page 5 of 9
			Attachment 2

Key principles for safe isolation must be followed:

- hazard identification;
- risk assessment and selection of procedure for isolation and decontamination;
- planning;
- isolation of equipment;
- testing effectiveness of isolation;
- decontamination;
- carrying out maintenance activity;
- reinstatement of plant;
- handover.

(Further guidance may be obtained from *The Safe Isolation of Plant and Equipment*, issued by the Oil Industry Advisory Committee of the UK Health and Safety Executive, ISBN 0 7176 0871 9.)

Chemicals Handling Inc., SHE Management Procedures
Procedure for Work Permits
Attachment 3 — Guidance on hot work permits

Reg. No: SHE-7.4	Revision: 1	Valid from:	Page 6 of 9
			Attachment 3

When issuing a hot work permit, consideration should be given to:

- ensuring that the area where the work will be carried out is clearly defined;
- removing the hazards (see above);
- minimizing the sources of hazard (e.g., removing flammable material from the area);
- testing the atmosphere before work starts;
- ensuring that activities liable to lead to the release of inflammable materials are prohibited (e.g., loading or discharge of tankers);
- ensuring that the atmosphere is re-tested at suitable intervals;
- ensuring that safety equipment, fire blankets, fire extinguishers, etc., are provided and used.

Hot work permits must be accepted by the person carrying out the work or that person's supervisor. The period of validity must be specified on the permit.

The permit is valid between the time limits stated on the form, with a maximum of eight hours. If the work is not started within two hours of the stated starting time or is interrupted for more than two hours, the permit becomes invalid.

In case of a gas or fire alarm, the permit becomes invalid and the work shall be interrupted immediately.

Copies of all permits in force must be kept in the control room. The permit must be signed off and returned to the control room on completion of the work.

Chemicals Handling Inc., SHE Management Procedures **Procedure for Work Permits** Attachment 4 — Guidance on issuing a confined space entry permit			
Reg. No: SHE-7.4	Revision: 1	Valid from:	Page 7 of 9
			Attachment 4

When issuing a confined space entry permit, the following aspects shall be considered:

- what are the hazards and risks?
- how can the space be isolated from sources of hazard (normally this will require physical isolation, e.g., by removal of connections)?
- hazardous materials which may be created during the work, i.e. vapours generated during cleaning or sludge removal, use of solvents, etc.;
- preparation of a plan showing all the isolations required;
- preparation of a rescue plan, in association with the appropriate trained staff, to rescue anyone from within the confined space should the atmosphere be hazardous;
- carrying out the isolations and checking their effectiveness;
- purging the confined space to remove hazardous materials;
- testing of the atmosphere by a qualified member of staff and recording the results of the tests and the oxygen content of the confined space;
- specifying the frequency at which the atmosphere should be re-tested.

If the confined space can be stated to be 'isolated and free from all sources of danger', a confined space entry permit may be issued. If it is not free of all sources of danger, additional precautions will be required, e.g., use of breathing apparatus, rescue line, arranging for attendant outside confined space, etc. These should be discussed with more senior management before a permit is issued.

The confined space entry permit should be accepted by the person entering the space or that person's supervisor.

The period of validity must be specified on the permit. The permits are valid between the time limits stated on the form, with a maximum of eight hours. If the work is not started within two hours of the stated starting time or is interrupted for more than two hours, the permit becomes invalid. In case of a gas or fire alarm, the permit becomes invalid and the work shall be interrupted immediately.

A copy of the permit must be kept in the control room. The permit should be displayed close to the point where entry is to be made. The permit must be signed off and returned to the control room on completion of the work.

The site manager must ensure that suitable staff are trained in rescue from confined spaces and that suitable equipment, self-contained breathing apparatus, air-line equipment, etc., *which can be used in confined spaces*, is available.

Chemicals Handling Inc., SHE Management Procedures
Procedure for Work Permits
Attachment 5 — Guidance on excavation permits

Reg. No: SHE-7.4	Revision: 1	Valid from:	Page 8 of 9
			Attachment 5

When issuing an excavation permit, the following aspects shall be considered:

• presence of underground electrical cables;
• presence of underground pipework and drains;
• possibility of encountering contaminated soil and groundwater;
• oxygen deficiency;
• ground collapse.

The excavation permit should be accepted by the person carrying out the excavation work, or that person's supervisor.

The period of validity must be specified on the permit. The permits are valid between the time limits stated on the form, with a maximum of eight hours. If the work is not started within two hours of the stated starting time or is interrupted for more than two hours, the permit becomes invalid. In case of a gas or fire alarm, the permit becomes invalid and the work shall be interrupted immediately.

A copy of the permit must be kept in the control room. The permit must be signed off and returned to the control room on completion of the work.

Chemicals Handling Inc., SHE Management Procedures
Procedure for Work Permits
Attachment 6 — Preparations before and after maintenance work: general guidelines

Reg. No: SHE-7.4	Revision: 1	Valid from:	Page 9 of 9
			Attachment 6

Equipment which is to receive maintenance work should be 'non-hazardous' – that is, free from dangerous chemicals, free from pressure, non-energized, etc.,

A standard procedure to prepare equipment containing a hazardous flammable chemical would contain a detailed instruction on the following steps:

- close all connections around the equipment;
- prove the isolations;
- drain the equipment (in a safe and environmentally sound way);
- flush the equipment (in a safe and environmentally sound way), e.g. with water;
- purge the equipment with nitrogen;
- introduce air and open the equipment.

Depending on the hazards, blinding of all or certain connections shall be carried out or a 'double block-and-bleed' arrangement could be used. For entry into vessels only blinding is allowed.

There should be a written instruction for each job covering the preparations for maintenance, including relevant sketches of the equipment and lists with closing and blinding. The documents shall be clearly signed when the actions (e.g., putting in a blind) have been performed.

When carrying out maintenance work on equipment involving rotating equipment, this must be isolated, safely locked or otherwise disconnected. In Chemicals Handling Inc. a system is used where the motor switch is disconnected and then locked physically in the non-energized position. Every person who will work on the equipment has their own lock.

After maintenance work it is equally important that the preparations for start-up are made in a safe and careful way following a similar procedure:

- flush the equipment (in a safe and environmentally sound way), e.g. with water;
- drain the equipment (in a safe and environmentally sound way);
- purge the equipment with nitrogen;
- check the tightness of the equipment;
- open connections around the equipment.

Before any equipment is taken into operation after maintenance, the system shall be physically checked in the field so that all valves which shall be closed (and sealed) are closed (and sealed) and vice versa.

Chemicals Handling Inc., SHE Management Procedures **Procedure for Contractors**				
Reg. No: SHE-7.5	Revision: 1	Valid from:		Page 1 of 2
Approved by:	Date:		Issued by:	
Distribution:				

Objective	The objective of this procedure is to regulate the conditions under which contractors may work for Chemicals Handling Inc. and state the applicable rules.
Scope	This procedure is applicable to all contractors. 'Contractors' is taken to mean: contractors, consultants, service personnel, transporters and other personnel (not employed by the company) conducting work for Chemicals Handling Inc.

Principles and methods	Only contractors who can demonstrate a good SHE performance shall be engaged by Chemicals Handling Inc. Contractors shall follow rules which apply to Chemicals Handling Inc. at the workplace. Contractors who flagrantly or repeatedly violate these rules shall be removed from the workplace.
	
	Purchase
	When purchasing contractor services the purchaser shall ensure that the contractor receives all relevant SHE information as a condition for an order. The contractor shall confirm that he has studied and accepted the SHE rules in his order acknowledgement. An insurance for responsibility for the contractor may be needed.
	
	Information/training
	Each individual person who works as a contractor shall be informed/trained in the SHE rules of Chemicals Handling Inc., before they can undertake any work. The information/training shall be documented and be signed off by the contractor.
	  The responsibility to give SHE information lies with the person placing the order (but can be done by another person).
	  Before placing the order the purchaser must be satisfied that the contractor can meet the SHE requirements (see the procedure for PURCHASING, SHE-6.1).
	
	Contact person
	Chemicals Handling Inc. shall have a contact person for every single contractor, to take responsibility for the contractor during the work. The person placing the order is responsible for this working in practice.
	  If there are many contractor persons simultaneously working for Chemicals Handling Inc. the contractor firm shall appoint one contact person for SHE matters.
	
	GENERAL RULES
	
	The contractor shall follow all general site rules of Chemicals Handling Inc. plus certain special rules. The applicable SHE procedures are listed under 'References'. The most important ones are:
	
	• GENERAL SITE RULES, SHE-7.1
	• WORK PERMITS, SHE-7.4
	• HANDLING OF CHEMICAL PRODUCTS, SHE-7.6
	• HANDLING OF WASTES, SHE-7.7
	• SOIL AND GROUNDWATER PROTECTION, SHE-7.8
	• NON-PROCESS HAZARDS, SHE-7.9

Chemicals Handling Inc., SHE Management Procedures **Procedure for Contractors**			
Reg. No: SHE-7.5	Revision: 1	Valid from:	Page 2 of 2

Principles and methods (cont'd)	**ADDITIONAL RULES** In addition to the general site rules above, the following rules also apply to contractors (in certain cases further additional rules can be issued): • All persons employed by contractors shall be registered when entering the industrial site of Chemicals Handling Inc. Checks should be made that relevant SHE information has been received. • All equipment used by the contractor shall comply with legislation and other official requirements. • Equipment must only be used within the limits for which it has been designed and approved. • Machinery and other equipment may only be used by qualified personnel. • Scaffolding may only be erected by qualified and approved personnel/firm. • All equipment shall be turned off after finishing the work or at the end of the day. • Storage of flammable/combustible material shall be in accordance with advice from Chemicals Handling Inc. • Erection of barracks, parking of vehicles, etc. must follow the rules issued by Chemicals Handling Inc.
Responsibility	The person placing the order is responsible for the contractor receiving the required information/training and observing the regulations which have been issued by Chemicals Handling Inc. Contractors are themselves responsible for occupational health matters within their own work. A specially appointed SHE contact person from the contractor can be responsible in some cases (to be specified in each individual case). The responsibility for co-ordination at the workplace lies with Chemicals Handling Inc.
References	This procedure refers to the following SHE procedures: SHE POLICY ORGANIZATION SHE LEGISLATION TRAINING PURCHASING GENERAL SITE RULES INSTRUCTIONS PLANT INTEGRITY AND MAINTENANCE WORK PERMITS HANDLING OF CHEMICAL PRODUCTS HANDLING OF WASTES SOIL AND GROUNDWATER PROTECTION NON-PROCESS HAZARDS MANAGEMENT OF CHANGE INVESTMENT PROJECTS SHE ROUNDS OCCUPATIONAL HEALTH EMERGENCY RESPONSE ACCIDENTS, INCIDENTS AND DISTURBANCES

Chemicals Handling Inc., SHE Management Procedures				
Procedure for Handling of Chemical Products				
Reg. No: SHE-7.6	Revision: 1	Valid from:		Page 1 of 2
Approved by:	Date:		Issued by:	
Distribution:				

Objective	This procedure describes the rules for handling the chemical products which are used within Chemicals Handling Inc.
Scope	This procedure applies to all types of activities and all sorts of chemical products.
Principles and methods	Chemicals Handling Inc. shall strive to work with those methods and equipment that pose the minimum risk to employees or the environment being exposed to dangerous chemical products.

General and references

Special permits are required for handling certain chemical products. This is regulated in the procedure for PRODUCT CONTROL, SHE-12.1. Classification, registration, investigation/assessment, labelling and safety and environmental information are regulated in the same procedure.

Purchasing of chemical products is regulated in the procedure for PURCHASING, SHE-6.1 along with the requirements on material safety data sheets and SHE assessment of the product.

Material safety data sheets

Material safety data sheets shall be obtained for all chemical products handled on site. Copies shall be kept:

- by the plant manager;
- on the plant, in a form accessible to all staff (including contractors);
- by others as specified.

A control system shall be in place for the issue, updating and withdrawal of material safety data sheets.

Handling

The handling of chemical products shall, where practicable, be carried out in closed systems. Where this is not possible the design shall be made with special consideration to minimizing the risks for SHE impact.

In addition to a good basic design (inherent safety) the SHE aspects shall be considered in training and instructions. The training is regulated in the procedure for TRAINING, SHE-5.2. There shall be separate detailed instructions for unloading, loading, storage and transport of chemical products (not accounted for here). See the procedure for INSTRUCTIONS, SHE-7.2. The applicable rules for handling, safety precautions and requirements regarding personal protection equipment shall be clearly stated in the instructions. There shall also be references to the relevant SHE information sheet or material safety data sheet for each chemical product in the operating instructions.

Personal protection equipment

According to Chemicals Handling Inc. policy, the goal is that employees and others shall be able to work without risk without having to use personal protection equipment. In certain cases it is not possible to achieve this goal, so appropriate personal protection equipment must be supplied by the company. Employees and other personnel are obliged to use the equipment where this is specified. In certain cases the safety instructions connected to the operations instructions require the use of special safety protection equipment. There shall be statements in the handling instructions about which activities require safety protection equipment.

Chemicals Handling Inc., SHE Management Procedures **Procedure for Handling of Chemical Products**			
Reg. No: SHE-7.6	Revision: 1	Valid from:	Page 2 of 2
Principles and methods (cont'd)	The responsible manager must never fail to act if an employee or contractor is violating the rules. Repeated violations of the rules can be reason for dismissal from the company of an employee or eviction from the site of a contractor.		
Responsibility	Overall responsibility for the safe handling of chemicals lies with each department manager. The line manager is responsible for ensuring that all material safety data sheets are available. The line manager is always responsible for specifying in the instructions the relevant handling requirements and the personal protection equipment to be used. The company is responsible for supplying adequate safety protection equipment through the line responsible manager. The supervisors and the department managers are primarily responsible for ensuring that the rules are followed. It is the responsibility of each employee to follow these rules and to keep the personal protection equipment in good condition.		
References	This procedure refers to the following SHE procedures: SHE POLICY SOIL AND GROUNDWATER PROTECTION SHE LEGISLATION NON-PROCESS HAZARDS SHE PERMITS MANUFACTURING METHODS/R&D WORK/ SCALE-UP HEALTH CARE RISK ANALYSIS TRAINING SHE ROUNDS GENERAL SITE RULES OCCUPATIONAL HEALTH INSTRUCTIONS ACCIDENTS, INCIDENTS AND DISTURBANCES CONTRACTORS PRODUCT CONTROL HANDLING OF WASTES		

Chemicals Handling Inc., SHE Management Procedures **Procedure for Handling of Wastes**				
Reg. No: SHE-7.7	Revision: 1	Valid from:		Page 1 of 3
Approved by:	Date:		Issued by:	
Distribution:				

Objective	The objective of this procedure is to give the framework for the relevant rules and instructions for handling wastes within Chemicals Handling Inc.
Scope	This procedure applies to all activities within Chemicals Handling Inc.

Principles and methods
 	
•
	The handling of wastes is regulated by a number of laws and other legislation. Chemicals Handling Inc.'s principles are as follows:
	
	• All legislation shall be fulfilled.
	• Effort should be directed to reduce the quantity of waste arising. This means that, at the design stages, technology and techniques which minimize the waste quantities and are adjusted to reuse and recycle, shall be applied.
	• The waste produced shall be handled in such a way to minimize the risks for humans and the environment, both internally within the company and externally, until the waste has been reused, deposited safely or neutralized by destruction.
	
	Priorities
	Waste shall be handled according to the hierarchy:
	
	1. Minimize the production of waste.
	2. Reuse.
	3. Recycle.
	4. Combustion.
	5. Deposition.
	
	Classification and source sorting
	Within Chemicals Handling Inc. there is a system for classifying, source sorting and handling of wastes according to Attachment 1. The treatment, collection method and procedure or contract that regulates the final disposal is stated for every type of waste.
	   In order to facilitate the source sorting, every workplace should have clearly marked containers for all categories of wastes unless there is another working system for delivering waste according to Attachment 1.
	   The storage shall be designed so that no risks to the working or external environment arise.
	   It is the responsibility of every employee to deliver the waste to the correct place. When in doubt, and on other occasions, the SHE function should be asked for advice.
	
	Declaration and labelling
	Declaration and labelling shall always follow legislative requirements.
	All wastes shall always be clearly labelled. The labelling of various categories is shown in Attachment 1.
	
	Records
	A full record shall be kept of all waste which is deposited to landfill.

Chemicals Handling Inc., SHE Management Procedures **Procedure for Handling of Wastes**			
Reg. No: SHE-7.7	Revision: 1	Valid from:	Page 2 of 3

Principles and methods (cont'd)	**Reporting** Waste reporting shall be made to the appropriate authority. The department manager is responsible for reporting waste (type, amount, way of disposal, etc.) to the SHE function, which summarizes the data and reports to the authority. See the procedure for SHE REPORTING, SHE-4.2. **Transport** When the waste is not transported by company personnel, the transport shall be handled by an approved external transport disposal company. **Contract** There shall be contracts for the final disposal of all types of wastes which can not be reused or recycled internally. **Training** All employees shall be trained in this procedure and other relevant instructions regarding waste. The responsibility for this lies with the line manager.
Responsibility	The responsibility for waste handling is split in the following way: The line organization where the waste is generated has prime responsibility, and then in turn the deliverer of waste and the transporter within the respective areas. The SHE function is responsible for co-ordination and expert advice. The transporter is responsible for adhering to the transport legislation and other relevant regulation. The SHE function can advise on these issues, when required.
References	This procedure refers to the following SHE procedures: SHE POLICY CONTRACTORS SHE OBJECTIVES AND ACTION PLANS HANDLING OF CHEMICAL PRODUCTS MANAGEMENT REVIEW SOIL AND GROUNDWATER PROTECTION SHE LEGISLATION MANUFACTURING METHODS/R&D WORK/SCALE-UP SHE PERMITS SHE AUDITS SHE EFFECTS/IMPACTS RISK ANALYSIS SHE REPORTING SHE ROUNDS TRAINING TRANSPORT EXTERNAL COMMUNICATION PRODUCT CONTROL INSTRUCTIONS

Chemicals Handling Inc., SHE Management Procedures
Procedure for Handling of Wastes
Attachment 1 — Example of classification and source sorting

Reg. No: SHE-7.7	Revision: 1	Valid from:	Page 3 of 3
			Attachment 1

Category	Principle for disposal	Collection method	Regulated in procedure/contract
Plastic	Recycle		
Steel, metal and cabling	Recycle		
Wood	Reuse Combustion	Container, labelled xx	
Paper	Recycle (Combustion)	Plastic box	
Other office waste	Combustion	Paper basket	
Fluorescent tubes	Special treatment		
Drums, empty	Reuse (Recycle)		
Hazardous waste	Special treatment	Specified procedure	Instruction 'Disposal of liquid hazardous waste'
Combustion industrial waste	Combustion	Specified procedure	
Computer equipment	Recycle		
Batteries	Special treatment	Battery boxes or sent to the electrical workshop	
Packings			

Chemicals Handling Inc., SHE Management Procedures **Procedure for Soil and Groundwater Protection**			
Reg. No: SHE-7.8	Revision: 1	Valid from:	Page 1 of 2
Approved by:	Date:		Issued by:
Distribution:			

Objective	The objective of this procedure is to give the basic rules and instructions for the protection of soil and groundwater within Chemicals Handling Inc.
Scope	This procedure applies to all activities within Chemicals Handling Inc.
Principles and methods	**General** As a basic principle there should be no releases from Chemicals Handling Inc.'s activities that could be harmful to the soil or groundwater. All spills of liquids or solid waste must be dealt with in an orderly manner. Draining of equipment, spills during sampling, etc., shall either go to closed systems or be dealt with in a controlled manner. **Strict discipline** It is the responsibility of every employee to deliver the spills to the correct place. Strict discipline is to be maintained. **Design measures** Areas susceptible to inadvertent spills should be constructed in such a way that spills cannot contaminate the soil or groundwater (for example, using a suitable hard surface, installing drainage and collection systems). **Control measures** The measures taken to avoid contamination shall include: • checking hard surfaces for tightness at least once per year – for example, by observing that rain water is contained and not leaking to the ground; • checking that drainage systems are clear (once every year); • checking that drainage systems, especially underground systems, are not leaking (once every three years, by pressure testing or inspection by fibre optics). **Existing contamination** There is a possibility of some soil contamination due to the previous activities at the site. The extent of this shall be investigated by sampling. After establishing the nature and the extent of this contamination, a programme will be introduced to control it in place or to take remedies to remove it. **Documentation of systems** Chemicals Handling Inc. shall have complete, updated documentation of all drainage and collection systems. **Final handling** Spills and wastes which are not handled in closed systems or suitably recovered shall be handled according to the procedure for HANDLING OF WASTES, SHE-7.7, including declaration, labelling and reporting. The final disposal of spills and wastes shall be in accordance with the same procedure. **Training** All employees and contractors shall be trained in this procedure and other relevant instructions. The responsibility for this lies with the line manager.

Chemicals Handling Inc., SHE Management Procedures				
Procedure for Soil and Groundwater Protection				
Reg. No: SHE-7.8	Revision: 1	Valid from:		Page 2 of 2
Principles and methods (cont'd)	**Preparedness for inadvertent spills** Chemicals Handling Inc. will handle major inadvertent spills according to the procedure for EMERGENCY RESPONSE, SHE-10.1, which defines the organizational resources, alarming, contacting the relevant authorities, etc. The company shall keep suitable equipment and material at the site to deal with any spills efficiently.			
Responsibility	The line organization has prime responsibility for handling spills. The SHE function can advise on these questions, when required.			
References	This procedure refers to the following SHE procedures: SHE POLICY INSTRUCTIONS SHE OBJECTIVES AND ACTION PLANS CONTRACTORS MANAGEMENT REVIEW HANDLING OF CHEMICAL PRODUCTS SHE LEGISLATION HANDLING OF WASTES SHE PERMITS SHE AUDITS SHE EFFECTS/IMPACTS RISK ANALYSIS SHE REPORTING SHE ROUNDS TRAINING TRANSPORT EXTERNAL COMMUNICATION			

Chemicals Handling Inc., SHE Management Procedures **Procedure for Non-Process Hazards**				
Reg. No: SHE-7.9	Revision: 1	Valid from:		Page 1 of 1
Approved by:	Date:		Issued by:	
Distribution:				

Objective	This procedure shall ensure that Chemicals Handling Inc. has the necessary and adequate instructions and rules for non-process hazards, and that they are implemented.
Scope	This procedure applies to the whole of Chemicals Handling Inc.
Principles and methods	**General** In Chemicals Handling Inc. there are a large number of hazards, which are general for most industries and not specific to the fact that the company handles chemicals in a process plant. It is as important that employees are protected from these hazards as well as from the more specific chemical hazards. **Specific instructions** In order to protect employees, contractors and others from the non-process hazards, certain work associated with these hazards is regulated. There are special instructions for: • work involving asbestos (identification, type of work, tools to be used, personal protection equipment, isolation of workplaces, measurements, authorization of work); • work at heights (type of work, use of safety harness and rope, use of ladders including checks, use of scaffolds including checks); • work involving biological hazards such as legionella (avoidance of exposure, use of personal protection equipment); • internal works transport (type of job, authorized persons, transport rules); • radiation hazards; • machine guarding (type of protection, checking procedure); • construction work.
Responsibility	The operations manager and project leaders have overall responsibility for adhering to these procedures.
References	This procedure refers to the following SHE procedures: SHE POLICY RISK ANALYSIS SHE LEGISLATION SHE ROUNDS HEALTH CARE OCCUPATIONAL HEALTH TRAINING ACCIDENTS, INCIDENTS AND DISTURBANCES GENERAL SITE RULES TRANSPORT INSTRUCTIONS CONTRACTORS

Chemicals Handling Inc., SHE Management Procedures **Procedure for Management of Change**				
Reg. No: SHE-7.10		Revision: 1	Valid from:	Page 1 of 5
Approved by:		Date:		Issued by:
Distribution:				

Objective	The objective of this procedure is to ensure that modifications and changes in the processes, plants and equipment of Chemicals Handling Inc. are carried out in such a manner that they do not increase risk for people, the environment or the facilities.
Scope	The procedure applies to all change and modification work, as defined below, within all activities of Chemicals Handling Inc. Larger projects with their own organization will handle SHE issues under the SHE procedure for INVESTMENT PROJECTS, SHE-8.2. Modifications within larger projects shall be handled in a similar way as in this procedure. Organizational changes with possible SHE effects are also included.
Principles and methods	A well structured work procedure, controlled by a special modification document with defined control steps and clearly defined responsible persons, shall ensure that all process aspects as well as safety and environmental aspects are covered and legislative requirements and the company internal rules are met. Associated with the modification procedure are other SHE procedures such as that for RISK ANALYSIS, SHE-9.2. **Definitions** A modification is a job which includes some change to the existing facilities or operations, including changes in the process, mechanical changes in the equipment, changes in programmable control systems and safety systems. Changes in the operating parameters which are outside normal variations will also be considered as modifications. The concept of modification includes both permanent and temporary changes. Attachment 1 gives a definition of 'modification' with a number of examples. **Work procedure** The work procedure for a job shall be as follows (see also Attachment 2). 1) The department manager (or supervisor) describes the proposal on a MOD(ification)-form giving a justification and a work order number. 2) The department manager approves the proposal for further processing and investigation, and decides who shall evaluate the SHE aspects. 3) An independent party assesses the SHE hazards and risks. (Where the appropriate procedure for RISK ANALYSIS, SHE-9.2, shall be followed.) The documentation is returned to the department manager following the investigation. Investigation reports, design documents, etc. (when relevant) are attached to the MOD-form. Report designations are stated on the MOD-form as a reference. 4) The department manager approves (or disapproves) the job. The department manager appoints a project leader to carry out the engineering based on the SHE assessment. 5) The project leader, in consultation with the department manager, decides what controls are necessary. 6) The project leader ensures that the necessary controls are carried out. On completion the project leader reports back to the department manager. 7) The department manager arranges for any changes to operating instructions and the training of staff. 8) A pre-start-up check is made to ensure that the installation is safe and the department manager signs the form for commissioning. *The modification is taken into operation* 9) The project manager ensures that all documentation is completed, revised and filed, and gives his authorization that all documentation is complete. The project manager is responsible for the MOD-form being filed in an official filing system.

Chemicals Handling Inc., SHE Management Procedures **Procedure for Management of Change**			
Reg. No: SHE-7.10	Revision: 1	Valid from:	Page 2 of 5

Principles and methods (cont'd)	**Controlling system – MOD-form** In order to have the sufficient control and uniformity throughout all modification work, a MOD-form shall accompany the whole work from idea/order through investigation/process design – engineering – construction – start-up – to final documentation. This form contains a number of control stations for approval by the relevant handling persons. The form should merely be seen as a check-list to ensure that all the important steps are taken care of.
Responsibility	The production manager is responsible for ensuring that modifications are carried out in accordance with this procedure. Besides that, the responsibility is defined above.
References	This procedure refers to the following SHE procedures: SHE POLICY SHE LEGISLATION SHE PERMITS SHE EFFECTS/IMPACTS TRAINING PURCHASING INSTRUCTIONS WORK PERMITS CONTRACTORS PROCESS SAFETY (SHE) INFORMATION ENGINEERING STANDARDS INVESTMENT PROJECTS MANUFACTURING METHODS/R&D WORK/SCALE-UP RISK ANALYSIS FIRE PROTECTION

Chemicals Handling Inc., SHE Management Procedures
Procedure for Management of Change
Attachment 1 – Definition of modification

Reg. No: SHE-7.10	Revision: 1	Valid from:	Page 3 of 5
			Attachment 1

Modifications can be both large and small, permanent or temporary. A change or an addition to equipment or a change in operation which can affect the manufacturing process or the control of this (under normal or abnormal conditions) through, for example, increased or decreased flow, temperature, pressure, composition, corrosivity, etc. is defined as a modification.

Mechanical
The concept of modification comprises all changes or additions to piping systems, machines or other equipment. Examples are the introduction, removal or change of a valve, non-return valve, restriction orifice, filter, heat exchanger, pump, gasket, insulation, etc.

Instrument and control systems
Changes of a shutdown limit or C_v of a control valve are examples of modifications to instruments. Most changes to a control system are considered as modifications.

Electrical
Over-current protection and change of cable quality are examples of modifications to electrical equipment.

Changes to operating parameters
Changes to operating parameters, which are outside normal process variations, are considered as modifications. The introduction of a new chemical (new raw material, new catalyst, new additional chemical, new chemical for cleaning, etc.) is also considered as a modification.

Temporary modification
Temporary changes shall be carried out according to the same procedure as permanent changes.

"Any modification should be designed, constructed, tested and maintained to the same standards as the original plant" (The official Flixborough investigation).

Changes which are not considered as modifications
Pure exchanges of equipment with identical equipment do not fall into the category of modification. Changes to equipment which is not in contact with the manufacturing facilities or equivalent and cannot influence the integrity of the process or the plant or the safety of personnel or the environment are not considered as modifications. However, changes to supporting structures are included in the procedure.

Chemicals Handling Inc., SHE Management Procedures
Procedure for Management of Change
Attachment 2—Work procedure, SHE aspects

Reg. No: SHE-7.10	Revision: 1	Valid from:	Page 4 of 5
			Attachment 2

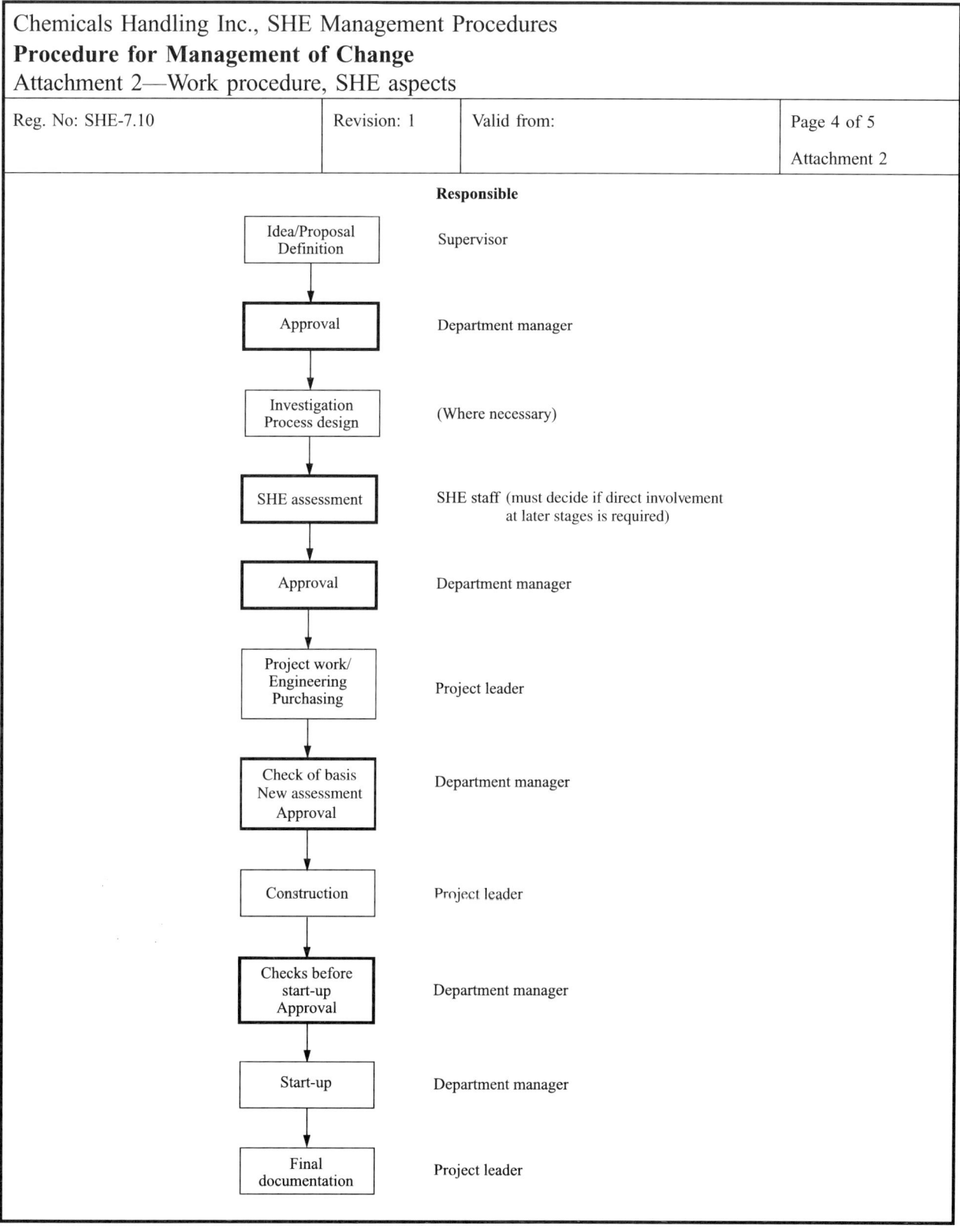

Responsible

Idea/Proposal Definition	Supervisor
Approval	Department manager
Investigation Process design	(Where necessary)
SHE assessment	SHE staff (must decide if direct involvement at later stages is required)
Approval	Department manager
Project work/ Engineering Purchasing	Project leader
Check of basis New assessment Approval	Department manager
Construction	Project leader
Checks before start-up Approval	Department manager
Start-up	Department manager
Final documentation	Project leader

Chemicals Handling Inc., SHE Management Procedures			
Procedure for Management of Change			
Attachment 3 – Pre-start-up checks			
Reg. No: SHE-7.10	Revision: 1	Valid From	Page 5 of 5
			Attachment 3

Each time after a modification has been made or when a project is nearing its completion, a pre-start-up check-out shall be undertaken.

Firstly, the person responsible for the modification or the project shall see that a check-out versus the engineering documentation and other relevant documentation is undertaken. Preferably this is to be carried out by somebody independent from the actual project organization. Possible deviations found at this check-out shall be noted on a 'but'-list and handed over to the operations staff. This document shall be filed with the other project or modification documentation.

Secondly, independent of the previous check-out, representatives from the operations shall check the installations before any commissioning work is started. This check shall be for general operability and maintainability as well as for checks versus drawings (e.g., P&I). This check should be filed in a binder for 'Operational check-outs'.

Chemicals Handling Inc., SHE Management Procedures **Procedure for Process Safety (SHE) Information**			
Reg. No: SHE-7.11	Revision: 1	Valid from:	Page 1 of 2
Approved by:	Date:		Issued by:
Distribution:			

Objective	This procedure shall ensure that Chemicals Handling Inc. has the necessary process safety information to maintain a high level of SHE standard.
Scope	This procedure applies to the whole of Chemicals Handling Inc.'s activities.

Principles and methods	**General**
	A complete and up-to-date set of engineering documentation is required to maintain a basis for a good level of performance in the SHE area. Important safety, health and environment documents for the safe performance of their jobs shall be available for all those requiring them.
	
	Categories of documents
	The following process safety (SHE) information shall be available (but not limited to):
	
	• Basic process information including:
	
	  – design basis;
	  – mass and energy balances;
	  – process flow diagrams (PFDs);
	  – general process equipment specifications;
	  – calculations of specific safety equipment such as safety valves and rupture discs;
	  – process description (including chemistry);
	  – process manual;
	  – engineering standards.
	
	• Detailed engineering documentation for all process and utility/service systems including:
	
	  – piping and instrument diagrams (P&IDs);
	  – general arrangement drawings (GAs);
	  – equipment specifications (e.g., pumps, heat exchangers, safety valves);
	  – electrical diagrams;
	  – earthing diagrams;
	  – instrument specifications;
	  – instrument loop drawings;
	  – software programs;
	  – interlock matrices;
	  – area classification drawings;
	  – isometrics;
	  – underground piping and cabling;
	  – layout drawings;
	  – plot plans;
	  – building drawings;
	  – foundation and structural steel drawings;
	  – specifications for effluent, waste disposal and noise.

Chemicals Handling Inc., SHE Management Procedures **Procedure for Process Safety (SHE) Information**				
Reg. No: SHE-7.11	Revision: 1	Valid from:		Page 2 of 2

Principles and methods (cont'd)	● Operating and maintenance documentation including:
	
	  – operating instructions;
	  – laboratory instructions;
	  – maintenance instructions;
	  – maintenance manuals;
	  – inspection files.
	
	● Other SHE documentation:
	
	  – material safety data sheets (MSDs);
	  – safety equipment documentation.
	
	Changes/modifications/updating
	All relevant safety (SHE) documentation shall be kept up to date. All relevant safety documentation shall be updated when performing modifications to the installations. Modifications shall follow the procedure for MANAGEMENT OF CHANGE, SHE-7.10. Key documents shall be approved by the department manager. Modifications shall be included on a 'master' drawing as sketches whilst waiting for the original drawing to be updated.
	
	Training
	Employees shall be trained in reading and understanding drawings and other important safety documents.
	
	Filing
	Originals of drawings and other important documentation shall be kept in central files in the main office building. They shall not be removed from there. There shall be a register for the easy retrieval of documents.

Responsibility	The operations manager has overall responsibility to maintain adherence to this procedure. Each manager is responsible for the documentation within his area.
References	This procedure refers to the following SHE procedures: SHE POLICY PLANT INTEGRITY AND MAINTENANCE MANAGEMENT REVIEW MANAGEMENT OF CHANGE SHE LEGISLATION ENGINEERING STANDARDS SHE PERMITS INVESTMENT PROJECTS SHE EFFECTS/IMPACTS MANUFACTURING METHODS/R&D WORK/SCALE-UP SHE REPORTING RISK ANALYSIS TRAINING OCCUPATIONAL HEALTH INSTRUCTIONS ENVIRONMENTAL CONTROL

Chemicals Handling Inc., SHE Management Procedures **Procedure for Engineering Standards**				
Reg. No: SHE-8.1	Revision: 1	Valid from:		Page 1 of 1
Approved by:	Date:		Issued by:	
Distribution:				

Objective	One of the best ways to design safe and reliable plants is to base the design on the data collected from the company and in the line of business. This should be documented in the form of engineering standards. The objective of this procedure is to define the importance of Chemicals Handling Inc.'s units being designed according to good engineering practice as documented in engineering standards, and to specify and follow these standards. The company standards shall be seen as additional to the rules and regulations on design and engineering which are given in official regulations, for example, on pressure vessels, tanks, piping.
Scope	This procedure applies to all project and maintenance activities within Chemicals Handling Inc. There are the following types of standards: • administrative standards, such as standards for drawings; • equipment and component standards, such as standards for piping and instrumentation; • instruction standards, such as standards for welding; • laboratory standards. There shall be a register of the relevant design and engineering standards.
Principles and methods	New projects shall follow the standards. At repair, modification or extension of existing units the relevant standards shall be used to the largest possible extent. Temporary installations shall also follow the standards. Standards can range from general guidelines to detailed descriptions of how to design a piece of equipment. In principle, standards shall be seen as mandatory. **Changes of standards** Changes of standards shall be regarded as a natural process since better designs and materials are continually appearing. The production manager shall approve a modification before it can be applied. The standard document shall be updated. **Deviation from standard** It is sometimes necessary to deviate from the present standard, particularly when working on equipment which does not follow the applicable standard. For work on systems standards, the earlier standard may be accepted if a competent and responsible person judges the installation to be equivalent to a design following the standard. Deviations from standards for temporary installations may only be made exceptionally. All deviations from applicable standards shall be approved by the department manager. **Auditing** Standards shall be audited once every two years and continuously when modifying the contents of a standard. The production manager is responsible for this.
Responsibility	The responsibility for complying with the standards lies with the person ordering the work. The responsibility for developing standards lies with the site manager.
References	This procedure refers to the following SHE procedures: SHE POLICY INVESTMENT PROJECTS PURCHASING TECHNOLOGY SALE PLANT INTEGRITY AND MAINTENANCE RISK ANALYSIS MANAGEMENT OF CHANGE

Chemicals Handling Inc., SHE Management Procedures **Procedure for Investment Projects**				
Reg. No: SHE-8.2	Revision: 1	Valid from:		Page 1 of 4
Approved by:	Date:		Issued by:	
Distribution:				

Objective	The objective of this procedure is to ensure that all phases of investment activities within Chemicals Handling Inc. are conducted in such a way that SHE aspects are treated in accordance with the company SHE policy.
Scope	This procedure applies to all investment activities within Chemicals Handling Inc. where there is a possibility for SHE effects. Those changes in activities which are not investment projects shall be handled from a SHE point of view according to the procedure for MANAGEMENT OF CHANGE, SHE-7.10.

Principles and methods	The basis for the safety, health and environmental aspects for the rest of the lifetime of the unit or product life cycle is laid during the initial projection of new equipment and units of manufacture of new products in existing units. There is a special procedure MANUFACTURING METHODS/R&D WORK/SCALE-UP, SHE-8.3 for regulating the manufacture of new products.
	   Coupled to all projects is the purchase of equipment, services, chemical products and other goods. Purchasing is regulated from a SHE point of view in the procedure for PURCHASING, SHE-6.1.
	   The implementation phase of an investment project often contains work introducing specific risks for safety, health and environment.
	
	Project organization
	For every project there shall always be a project manager with overall, total responsibility (often the site manager or a department manager) and a project leader (often a department manager or a supervisor or an external resource). The project leader leads the current activities of the project.
	   Additional support and control resources may be needed depending on the size of the project.
	   Normally there should be a project group including representatives of the employees (e.g., safety representatives) and of the SHE function.
	
	SHE review of projects
	The SHE review of a project shall generally be carried out by the SHE function but for major projects may be made by another competent person who shall be independent from the designer/engineer. The review shall be made according to the procedure for RISK ANALYSIS, SHE-9.2 and SHE aspects shall be investigated according to the procedure for SHE EFFECTS/IMPACTS, SHE-4.1. The result should be compared with the SHE policy and the SHE objectives of the company.
	
	Work procedure
	A simple work procedure for those project activities which require SHE reviews is outlined below and shown in Attachment 1. There should be a special Project/Modification form used for all projects to document that all SHE aspects have been considered.
	
	1) Pre-study, design basis
	The objectives and the framework of the project shall be defined in this phase. The SHE aspects shall be considered and be included in the project description.
	
	2) Process or basic design
	The SHE aspects shall be thoroughly investigated during this phase and the conditions be specified in detail. The SHE work shall follow the requirements in the procedure for RISK ANALYSIS, SHE-9.2, which means that a formal risk analysis, at least of a basic nature, shall be carried out.

Chemicals Handling Inc., SHE Management Procedures **Procedure for Investment Projects**				
Reg. No: SHE-8.2		Revision: 1	Valid from:	Page 2 of 4

| **Principles and methods (cont'd)** | *3) Detailed engineering*
All flow diagrams (P&I), manufacturing, general arrangement and construction drawings shall be produced during this phase. The engineering shall follow the Chemicals Handling Inc. procedure for ENGINEERING STANDARDS, SHE-8.1 (as well as all legislative requirements for engineering and construction). Deviations from a standard can only be made with the site manager's approval or in some cases the production manager's, in consultation with the project leader or the customer. During the detailed engineering, representatives from the customer – the department personnel – shall be given opportunity to influence issues of operability, maintainability and SHE.
   A detailed technical SHE review shall be made during this phase, when detailed material has been produced. The procedure for RISK ANALYSIS, SHE-9.2, regulates this.

4) Purchasing
The SHE requirements shall be carefully considered and specified in purchasing work. The procedure for PURCHASING, SHE-6.1, regulates how the SHE aspects shall be considered in purchasing.

5) Construction/implementation
This phase contains mostly physical completion and testing of equipment (usually without operational material in the systems). This often means an increased risk of injuries and damage, e.g. falling, crushing, cutting and electrical damage. In order to fulfil the rules of Chemicals Handling Inc. for workers' protection/occupational health, resources shall be allocated by the project leader. The procedure for CONTRACTORS, SHE-7.5, is valid for contractors. The fulfillment of the procedure for WORK PERMITS, SHE-7.4, is also vitally important. It is the responsibility of the project leader to see that these rules are followed. SHE rounds shall be made on construction sites.
   A special formal occupational health plan is needed for bigger projects containing, among other things, general rules of order and safety, the organization of the safety work and any special risks.
   Before the hazardous materials are introduced into the equipment, the department manager shall ensure that a 'pre-start-up safety review' is conducted. The review shall ensure that:

• all safety, health and environmental issues identified during the course of the project have been addressed and executed in a suitable manner;
• there is safe access to and from the workplace;
• operating procedures are available;
• staff have been trained;
• safety equipment is available and has been tested;
• the emergency procedures have been reviewed and that, where changes have been made, these have been practiced;
• procedures for routine testing needed for occupational health, environmental protection and safety have been prepared.

   On completion of this review the department manager shall record this in the SHE dossier indicating any conditions attached to the authorization for start-up.

6) Start-up/commissioning
Normally the customer (the receiving department manager) is responsible for start-up/commissioning, but it could also be done under the supervision and responsibility of the project leader. Regardless of who is responsible, it is of utmost importance that the personnel who will have responsibility for the operation of this equipment get adequate training. The person responsible for start-up is responsible for organizing this training. |

Chemicals Handling Inc., SHE Management Procedures **Procedure for Investment Projects**				
Reg. No: SHE-8.2	Revision: 1	Valid from:		Page 3 of 4
Principles and methods (cont'd)	From a SHE point of view, it is important that new equipment is function tested before being taken into normal operation. This shall in relevant cases be done with non-hazardous media or in other safe ways according to special procedures. Special rules and orders must be set up if start-up is carried out simultaneously with ongoing construction. The various responsibilities in this hand-over/start-up phase shall be regulated in a clear, unambiguous way, and shall be in writing for complicated situations. **Documentation** All project material of a technical nature shall be assembled in special binders. The documentation shall be available in the company or department filing system. The project leader is responsible for handing over the documentation. **Authority contacts/approval** In order to assist the project leader and be able to manage authority contacts with high internal competence via established contact avenues, the SHE function shall be the authority's contact in the SHE area.			
Responsibility	The primary responsibility for taking into account the SHE aspects in all phases of the project lies with the project leader and ultimately with the project manager. The project leader shall ensure that there is adequate competence within the project and that SHE reviews, authority approvals, etc., are carried out according to this procedure. The responsibility of co-ordinating all activities (including external workforces) during the construction/implementation phase lies with Chemicals Handling Inc. and, in this case, the project leader.			
References	This procedure refers to the following SHE procedures: SHE POLICY MANAGEMENT OF CHANGE ORGANIZATION PROCESS SAFETY (SHE) INFORMATION SHE LEGISLATION ENGINEERING STANDARDS SHE PERMITS MANUFACTURING METHODS/R&D WORK/SCALE-UP SHE EFFECTS/IMPACTS RISK ANALYSIS TRAINING SHE ROUNDS PURCHASING ACCIDENTS INSTRUCTIONS INCIDENTS AND DISTURBANCES WORK PERMITS FIRE PROTECTION CONTRACTORS TRANSPORT HANDLING OF WASTES			

Chemicals Handling Inc., SHE Management Procedures
Procedure for Investment Projects
Attachment 1 – Work procedure with regard to SHE aspects

Reg. No: SHE-8.2	Revision: 1	Valid from:	Page 4 of 4
			Attachment 1

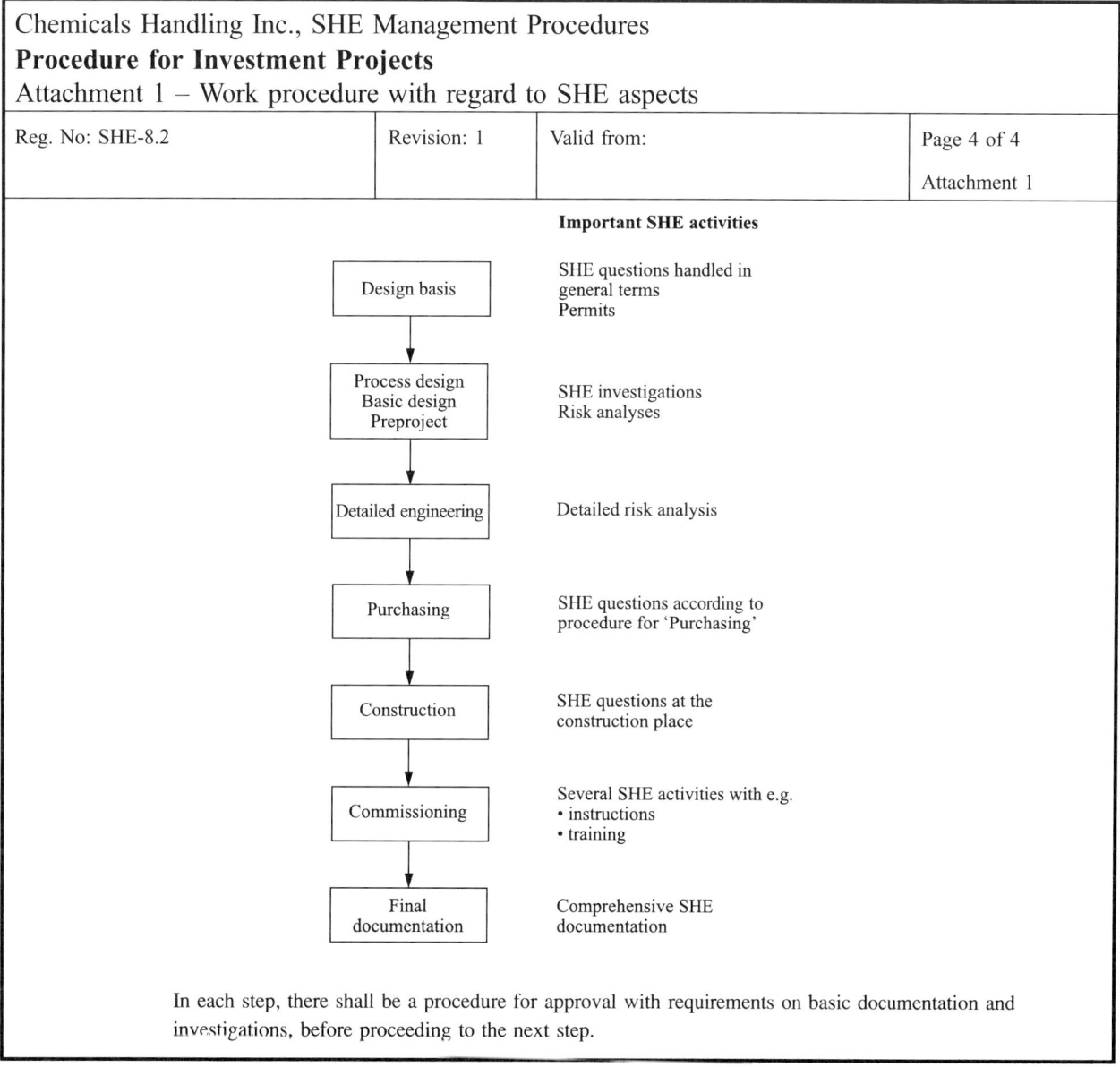

Important SHE activities

Design basis	SHE questions handled in general terms Permits
Process design Basic design Preproject	SHE investigations Risk analyses
Detailed engineering	Detailed risk analysis
Purchasing	SHE questions according to procedure for 'Purchasing'
Construction	SHE questions at the construction place
Commissioning	Several SHE activities with e.g. • instructions • training
Final documentation	Comprehensive SHE documentation

In each step, there shall be a procedure for approval with requirements on basic documentation and investigations, before proceeding to the next step.

Chemicals Handling Inc., SHE Management Procedures				
Procedure for Manufacturing Methods/R&D Work/Scale-up				
Reg. No: SHE-8.3		Revision: 1	Valid from:	Page 1 of 3
Approved by:		Date:		Issued by:
Distribution:				

Objective	The objective of this procedure is to ensure that the selection and development of new products, and their manufacturing methods or changes following development and scale-up work, will be carried out in a structured way with due consideration to SHE aspects in accordance with Chemicals Handling Inc.'s SHE policy.
Scope	This procedure applies to all products and manufacturing methods.
Principles and methods	Effective consideration can be made of the safety, health and environmental aspects of the manufacturing of a product during the early stages of development. In the later stages, time schedules seldom allow any fundamental changes to the selection of methods or chemicals. A well-structured work procedure with defined control steps for safety/health/environment forms the basis for developing the method. This procedure defines the requirements for carrying out SHE reviews (with scope and methods) at defined points of a project, after which formal approval by an authorized person is required before proceeding to the next step. All SHE aspects shall be evaluated and considered in the decisions regarding: • raw material selection and energy consumption; • SHE risks associated with operations; • product properties and waste problems. The SHE review will need to be repeated and approved whenever a chemical or manufacturing method is changed. **Work procedure** The following illustrates a typical work procedure (see also Attachment 1). *1) Idea/proposal/definition* The origin of a project can come from several sources, both internal and external. An overall definition of the project, including certain considerations of SHE issues, is stated. A 'project leader' is appointed. The site manager's approval is required. *2) Survey and selection of manufacturing method* The project leader shall undertake this task paying great consideration to SHE aspects. Risks shall be evaluated with regard to reactivity, fire, explosion, toxicity and environmental properties during manufacturing, use of and disposal of the product. For new products, a life cycle analysis/consideration shall be performed, if relevant. This survey shall be documented. The site manager shall approve it. *3) Method development (if relevant)* The method development is performed by the project leader. Before each experiment in the laboratory, a safety control with, among other things, a literature search shall be carried out covering the chemicals and the reaction synthesis to be used. Minor changes in the synthesis route can be approved by the production manager. When the method development is approaching its completion, a basis for SHE aspects shall be prepared by the project leader. The contents of such a document are defined in a special instruction.

Chemicals Handling Inc., SHE Management Procedures			
Procedure for Manufacturing Methods/R&D Work/Scale-up			
Reg. No: SHE-8.3	Revision: 1	Valid from:	Page 2 of 3

Principles and methods (cont'd)	*4) SHE review* A SHE review shall be performed by a group of people made up of the SHE manager (chairman), R&D responsible person, the project leader, other associated members and external specialists in areas such as environment, toxicology if necessary. Normally the group makes a simple evaluation (from written documentation) but more comprehensive risk analyses are necessary in certain cases. The basis for the review consists of the earlier mentioned survey/summary, instructions for operations. The project leader is responsible for assembling all the necessary material. *5) Full-scale operation* Before full-scale operation may commence, the production manager shall give approval for the run. In those cases where the SHE review has led to recommendations, these shall be carried out unless clear justification why action is not required can be given. In addition to having instructions for the run, the personnel concerned shall be adequately trained. The production manager is responsible. A decision to omit pilot-scale operation (where it could be relevant) can be taken by the site manager on recommendation from the SHE review group. *6) Operation in pilot plant* In those cases where a pilot plant operation precedes full-scale operation, the same procedure as for full-scale operation shall be applied. The experiences from the pilot plant operation shall be evaluated and a renewed SHE review shall be carried out before the step to full-scale operation may be taken. The production manager is responsible. **Controlling system** To achieve an adequate level of control throughout this procedure there shall be a form following the work from idea/proposal through the laboratory/R&D work, in which the requirements on reviews, documentation, training and formal approvals will be stated.
Responsibility	Before the start of development of a 'new product' or manufacturing method the responsibility for initiating and leading the work up to and including the documentation according to this procedure, lies with the site manager. Responsibility is also specified in the above procedure.
References	This procedure refers to the following SHE procedures: SHE POLICY · HANDLING OF WASTES SHE LEGISLATION · MANAGEMENT OF CHANGE SHE EFFECTS/IMPACTS · INVESTMENT PROJECTS TRAINING · RISK ANALYSIS PURCHASING · OCCUPATIONAL HEALTH INSTRUCTIONS · PRODUCT CONTROL HANDLING OF CHEMICAL PRODUCTS

Chemicals Handling Inc., SHE Management Procedures
Procedure for Manufacturing Methods/R&D Work/Scale-up
Attachment 1 – Work procedure with regard to SHE aspects

Reg. No: SHE-8.3	Revision: 1	Valid from:	Page 3 of 3
			Attachment 1

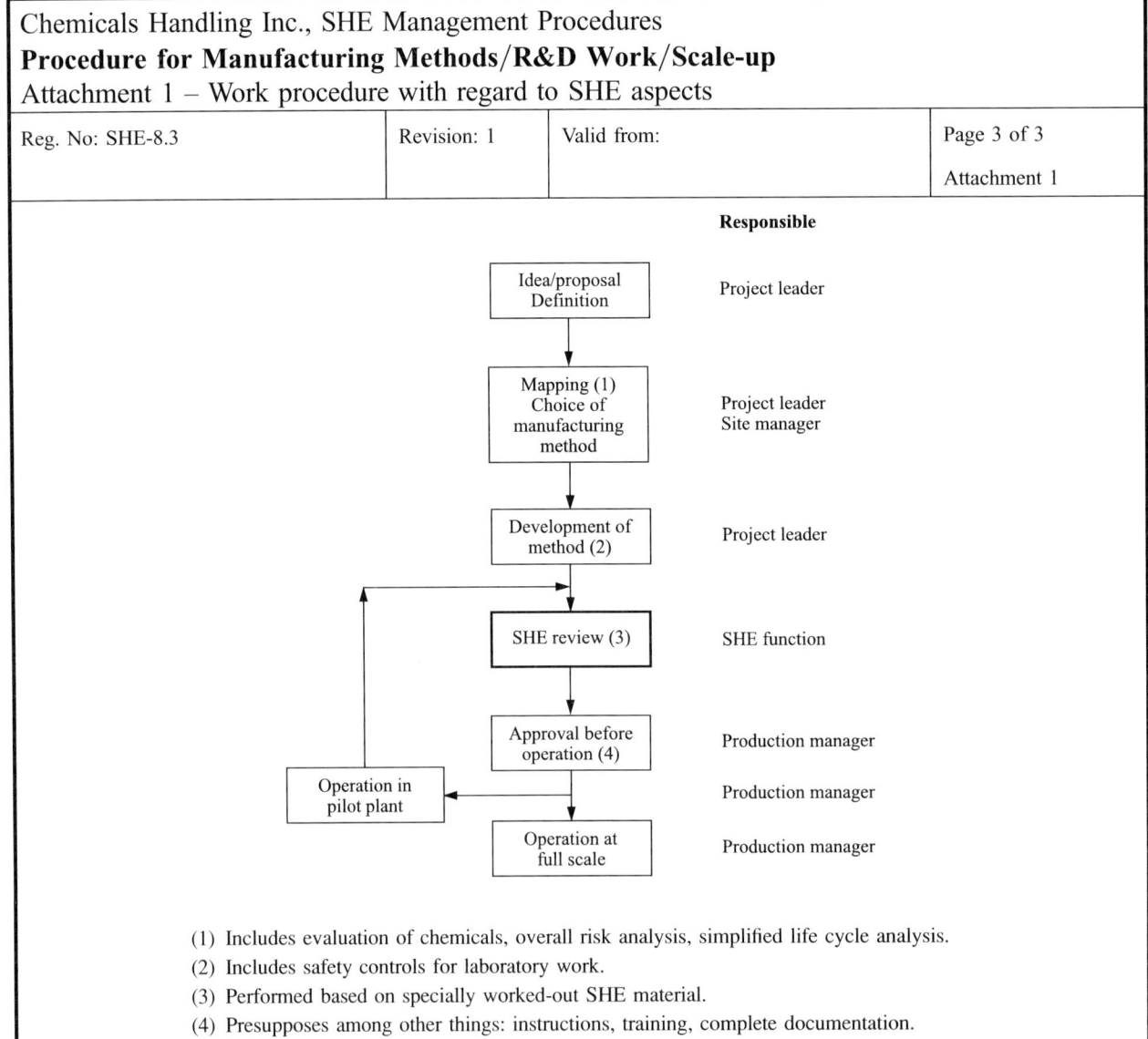

Responsible

Idea/proposal Definition	Project leader
Mapping (1) Choice of manufacturing method	Project leader Site manager
Development of method (2)	Project leader
SHE review (3)	SHE function
Approval before operation (4)	Production manager
Operation in pilot plant	Production manager
Operation at full scale	Production manager

(1) Includes evaluation of chemicals, overall risk analysis, simplified life cycle analysis.

(2) Includes safety controls for laboratory work.

(3) Performed based on specially worked-out SHE material.

(4) Presupposes among other things: instructions, training, complete documentation.

Chemicals Handling Inc., SHE Management Procedures **Procedure for Technology Sale**					
Reg. No: SHE-8.4		Revision: 1	Valid from:		Page 1 of 1
Approved by:		Date:		Issued by:	
Distribution:					

Objective	The objective of this procedure is to define the principles that Chemicals Handling Inc. shall apply during the transfer of its technology to another company to continue to guarantee the responsible use of company technology and the products based on it.
Scope	This procedure applies to all types of technology transfer (sale, licensing, toll manufacturing, etc.) relating to Chemicals Handling Inc.'s activities.

Principles and methods	Whenever technology is sold the objective shall be to ensure that SHE standards equivalent to those of Chemicals Handling Inc. are applied.
	   Chemicals Handling Inc. shall follow the advice on technology sale in the Responsible Care programme and in the International Chamber of Commerce (ICC) 16 point programme.
	   The sale of technology shall be made only to companies which can be judged to take full responsibility for SHE. Chemicals Handling Inc. shall gather information on how the customer plans to utilize the technology. This should be evaluated in connection with the technology sale.
	   The technology sale shall be made with active support in SHE matters from Chemicals Handling Inc. to the customer. This means the transfer of all relevant SHE information, training of the customer personnel, and controls put in place to ensure that SHE matters have been clearly understood and will be correctly handled.
	   The customer shall, by contract, commit himself to utilize the technology in a responsible way. Serious deviation from this shall give reason for re-negotiation of the transfer.
	   A document shall be drawn up which shows the considerations and actions that have been taken.

Responsibility	The managing director of Chemicals Handling Inc. is responsible for following this procedure.
References	This procedure refers to the following SHE procedures: SHE POLICY HANDLING OF CHEMICAL PRODUCTS SHE LEGISLATION HANDLING OF WASTES SHE EFFECTS/IMPACTS RISK ANALYSIS TRAINING PRODUCT CONTROL INSTRUCTIONS

Chemicals Handling Inc., SHE Management Procedures **Procedure for SHE Audits**				
Reg. No: SHE-9.1	Revision: 1	Valid from:		Page 1 of 2
Approved by:	Date:		Issued by:	
Distribution:				

Objective	The objective of this procedure is to define the requirements of regular audits to ensure that Chemicals Handling Inc. has an adequate SHE management system, and that it works in practice. This shall be accomplished with the aid of SHE audits according to a special form. The audits and the actions originating from them shall have the objective of ensuring that the whole organization is in good shape regarding safety, health and environment, and that a deterioration or too low a standard will be detected and corrected.
Scope	This procedure applies to all units within Chemicals Handling Inc. SHE auditing shall be conducted on two levels: 1) The whole company shall be audited in detail by wholly or partly independent auditors. 2) Each department shall undergo an audit, which shall be conducted with department personnel under the guidance of an independent company person, or external person.

Principles and methods	A SHE audit is a control of mainly the management system, to ensure that a business is run in a safe way, taking into consideration among other things:
	
	• information and communication;
	• SHE procedures;
	• training;
	• SHE issues in the workplace;
	• attitudes.
	
	    The focus shall be on programmes and systems rather than on detailed weaknesses. Purely technical aspects are not covered by this type of audit.
	    The auditing method to be used could be based on the simple model which is described in the Chemical Industries Association *Guidance on Safety, Occupational Health and Environmental Protection Auditing* or the Association of Swedish Chemical Industries guideline *SHE-AUDIT, A Guideline for Internal Auditing of SAFETY/HEALTH/ENVIRONMENT, 1996*. Chemicals Handling Inc. has made a company adjusted version of the auditing performance guides.
	The practical performance of the audit in the form of:
	
	• planning and preparation;
	• performance at site (including deeper analysis of selected systems);
	• reporting;
	• action plan and follow-up;
	
	shall be further defined. This is best done based on the description in the guideline.
	
	Audits
	
	Level 1 audit, the whole company
	The audit shall be conducted by a team of experienced persons with different and mixed backgrounds. The team interviews employees from different levels in the company, reviews various documents and makes a site visit.
	    Based on the interviews, document reviews and site visit, an evaluation of a large number of activities is made.

Chemicals Handling Inc., SHE Management Procedures **Procedure for SHE Audits**				
Reg. No: SHE-9.1	Revision: 1	Valid from:		Page 2 of 2
Principles and methods (cont'd)	*Level 2 audit, department level* In this case the audit shall be conducted with the personnel of the department, and evaluated according to the same performance guide as above. **Audit team** The audit team for auditing the whole company shall consist of a number of suitable persons, who are not involved in Chemicals Handling Inc.'s activities. These can be totally independent of the company or corporate employees or a mixture. The competence requirements of the team members regarding safety, health and environment as well as technical matters, legislative matters and management systems shall be established by the audit team leader and company management team, taking into account the requirements of the standards. The requirements of the team leader shall be specially considered. Generally speaking, persons with broad management competence and relevant speciality competence are needed. The audit at the departmental level shall mainly be conducted by the personnel from other departments on site and a selection of representatives from all levels within the department — under the guidance of an independent person from the SHE function. **Audit intervals** A level 1 audit shall be conducted every three to five years, or more frequently if required. The company management team shall agree on the frequency. A level 2 audit shall be conducted once per year. **Documentation** The audit shall be documented in writing. A comprehensive report is required for level 1, with references to reviewed material, identified weaknesses and recommendations. Only a short summary is needed for level 2, with the rating and, where relevant, recommendations for actions. **Follow-up** The audit report shall be distributed to the responsible manager (site manager or managing director for level 1, department manager for level 2). Based on this an action plan shall be developed in the Chemicals Handling Inc.'s management team's respective department meetings.			
Responsibility	Overall responsibility for ensuring that audits are conducted lies with the site manager/managing director. Department managers are responsible for ensuring that departmental audits are conducted. The SHE function is responsible for developing the auditing methods and for controlling the quality of the performed audits. The site manager and department managers are responsible for ensuring that all actions are followed up.			
References	This procedure refers to the following SHE procedures: ALL			

Chemicals Handling Inc., SHE Management Procedures **Procedure for Risk Analysis**				
Reg. No: SHE-9.2	Revision: 1	Valid from:		Page 1 of 2
Approved by:	Date:		Issued by:	
Distribution:				

Objective	The objective of this procedure is to state the requirements for the systematic use of technical SHE reviews, risk analyses. By using the most suitable analysis method for each individual case, safety can be optimized; i.e., a good balance between resources used and results obtained can be achieved.
Scope	Risk analyses shall in principle be carried out for all types of manufacturing, development, maintenance, transport work, etc. All projects and modifications shall also be subject to a risk analysis. Risk analyses may be renewed for special reasons (authority requirement, increased number of incidents, special company decision, etc.). A new risk analysis must be performed when changing systems, equipment, etc. Systems which cannot influence the safety or environmental aspects may be excluded from the review requirement.
Principles and methods	A sound hazard identification is the basis of all safety work. There are formal requirements on risk analysis in various legislation, such as the national applications of the Seveso directive. These legislative requirements form the basis for the risk analyses to be carried out within Chemicals Handling Inc. Risk analysis is a natural element of all projects and modification work. The work procedure, responsibility, etc. for these are regulated in special procedures – MANAGEMENT OF CHANGE, SHE-7.10 and INVESTMENT PROJECTS, SHE-8.2. All risk analysis work consists of these main steps: 1) Hazard identification. 2) Assessment of probability and consequence of the risk. 3) Evaluation of the risk. 4) Weighing the risk against a risk criterion (the company's tolerable risk levels need to be defined). **Preliminary risk analysis/classification** The first step in the review is to determine the risk level by use of a preliminary risk analysis. Depending on how the object (the activity or the process/modification) is classified, a detailed review, a semi-detailed review or no review at all, is required. The classification is semi-quantitative. A risk analysis shall normally be summarized in writing and contain a clear summary of the recommended actions. The method is based on assessments, by experienced personnel, of the probability and consequence of possible hazard scenarios. In Chemicals Handling Inc.'s application of this method, three categories of consequences are evaluated: ● human life and health; ● environment; ● property. **Performance of risk analysis** One basic principle for risk analyses is that they should be carried out by an independent party rather than the persons who designed the system or proposed the modification. For larger projects or modifications, the work should be carried out by a risk analysis group, comprising representatives from all relevant disciplines.

Chemicals Handling Inc., SHE Management Procedures **Procedure for Risk Analysis**			
Reg. No: SHE-9.2	Revision: 1	Valid from:	Page 2 of 2

Principles and methods (cont'd)	The person leading the risk analysis should be trained and experienced in the conduct of the relevant techniques (this may require the use of outside consultants). A record of the training and experiences of the risk analyst shall be kept by the SHE function. **Updating** Risk analyses shall be reviewed to check their validity at an interval no greater than once every 5 years. **Documentation** All risk analyses shall be documented in such a way that they could be reviewed on demand. In its simplest form the analysis is a judgement/statement by an authorized person, who by his or her signature on the 'Modification form' (according to the procedure for MANAGEMENT OF CHANGE, (SHE-8.4) or on the project documentation (according to the procedure for INVESTMENT PROJECTS, SHE-8.3)), approves an installation or a manufacturing process. In most cases there should be some form of written report, showing the objective, scope and result of the performed risk analysis. Such documentation shall be filed centrally by the SHE function. **Exemptions from the review requirements** Systems which cannot affect safety or environment may be exempted from the review requirements. The site manager or the department manager may also make exemptions where a risk analysis would not contribute anything of value. This decision shall be recorded in writing and with a justification.

Responsibility	The responsibility for having risk analyses carried out lies with the department manager, or the project leader in the case of projects. The responsibility for implementing the results of risk analyses including the recommendations also lies with the manager of the unit or project leader. The quality of the risk analysis is the responsibility of the next higher line manager.
	   The SHE function is responsible for ensuring that suitably qualified staff, either company employees or consultants, are available.
	   In the case of smaller jobs, the SHE function may make the risk analysis alone or together with someone else. Personnel within the production units or equivalent departments can do this once they have acquired adequate competence in carrying out risk analysis. When in doubt, the SHE function will decide how the analysis shall be performed.

References	This procedure refers to the following SHE procedures: SHE POLICY MANAGEMENT OF CHANGE SHE LEGISLATION INVESTMENT PROJECTS SHE EFFECTS/IMPACTS MANUFACTURING METHODS/R&D WORK/SCALE-UP

Chemicals Handling Inc., SHE Management Procedures **Procedure for SHE Rounds**				
Reg. No: SHE-9.3	Revision: 1	Valid from:		Page 1 of 2
Approved by:	Date:		Issued by:	
Distribution:				

Objective	The objective of this procedure is to define the rules for SHE rounds. The objective of SHE rounds is to identify and correct deficiencies in safety, occupational health and environment, by active co-operation between the company and the employees. In this way, work injuries and environmental damages can be minimized and hopefully totally eliminated.
Scope	This procedure applies to all departments within Chemicals Handling Inc. The rounds shall cover matters which could influence both the occupational health of the employees and the environment.
Principles and methods	Safety to health and the environment are created together through active involvement in the workplace. One of the indicators of commitment to SHE is the orderliness and housekeeping in plants, laboratories, workshops, etc. SHE rounds are one of the tools for controlling the safety status of the workplace. **Method** SHE rounds shall be given ample time. Their contents should be planned in advance subject to agreement between the department manager and the safety representative. 　SHE rounds should alternate between general SHE rounds, where all aspects are covered, and rounds with a special theme such as falling and tripping risks, exposure to chemicals, releases to the environment and transport risks. Factors such as incidents, work accidents and psycho-social issues should be dealt with during the rounds. 　SHE rounds should at least partly be a system review, i.e. checking that procedures exist and are followed. **Planning** Each unit or department shall plan its own SHE rounds for the coming year. At the latest, this planning should be ready by December. The programme shall be sent to the SHE committee and the main safety representative. **Frequency** Major SHE rounds shall be carried out four times per year in the production department and twice per year in all other departments. In addition to these official rounds, further rounds may be carried out after agreement between the respective manager and safety representative or employees. **Participants** Permanent members of the SHE round group shall include the acting department or section supervisor and the safety representative. Besides these, the department manager shall participate in at least two of the four rounds per year. The SHE function and the main safety representative of Chemicals Handling Inc. should also participate in most rounds. 　Representatives from the company health care area should participate on special demand. **Authority of the group** The SHE round group's decisions on actions within the occupational health area, which are contained within the current budget, shall be considered as binding. More costly actions will always need the approval of the department manager or the site manager. **Documentation** SHE rounds shall be minuted on a special form. The company representative of the group is responsible for this. The minutes shall be distributed to the participants, the department manager, the SHE function, the main safety representative and all others concerned.

Chemicals Handling Inc., SHE Management Procedures **Procedure for SHE Rounds**			
Reg. No: SHE-9.3	Revision: 1	Valid from:	Page 2 of 2
Principles and methods (cont'd)	**Follow-up** Reporting and follow-up shall be done departmentally and in Chemicals Handling Inc.'s SHE committee. The relevant department manager is responsible for reporting.		
Responsibility	The responsibility for ensuring that SHE rounds are carried out, reported and the results are implemented in practice lies with the department managers.		
References	This procedure refers to the following SHE procedures: SHE POLICY OCCUPATIONAL HEALTH GENERAL SITE RULES ENVIRONMENTAL CONTROL INSTRUCTIONS ACCIDENTS PLANT INTEGRITY AND MAINTENANCE INCIDENTS AND DISTURBANCES WORK PERMITS FIRE PROTECTION HANDLING OF CHEMICAL PRODUCTS PRODUCT CONTROL HANDLING OF WASTES		

Chemicals Handling Inc., SHE Management Procedures **Procedure for Occupational Health**						
Reg. No: SHE-9.4		Revision: 1	Valid from:			Page 1 of 2
Approved by:		Date:		Issued by:		
Distribution:						

Objective	The objective of this procedure is to define the scope, methods and follow-up of occupational health surveys.
Scope	This procedure applies to all activities within Chemicals Handling Inc.
Principles and methods	Certain chemicals which can be dangerous to health in very small amounts are handled at Chemicals Handling Inc. Besides that there are a small number of cases of working conditions which could create distress in other ways, such as a noisy environment and ergonomic problems. Chemicals Handling Inc.'s goal is to ensure that its employees are not exposed to dangerous doses of chemicals or other occupational health factors. Occupational exposure limits shall be kept within a good margin. Where such limits are missing, company internal limit values should be developed based upon comparison with similar chemicals and which, by a sound margin, could be considered as non-dangerous. In order to check that this goal is achieved, measurements in the working environment shall be taken regularly. **Inventory and measurement programme** The managing director shall ensure that the occupational health risks are assessed by a competent person. Where necessary an external consultant may be employed. A general inventory of the exposure risks in the workplace shall be made. Measurement programmes shall be developed for each department. The general scope and contents of measurement programmes shall be approved by the SHE committee/management group of Chemicals Handling Inc. Further measurements and other investigations should be added as required. Measurements shall be performed when using a chemical for the first time in the production or, in certain cases, in other departments. Thereafter recurrent measurements shall be made, at a frequency based on the results of the previous measurements. Measurements for unit operations, cleaning operations or maintenance shall be included in the inventory of the occupational health situation. **Evaluation and actions** The measurement results shall be evaluated by a person especially trained in measurement techniques. If the measurement shows that the exposure exceeds the Threshold Limit Value, action must be taken to decrease the exposure. Technical measures in the processes and/or the equipment shall be given priority over the use of personal protection equipment. The measurement shall then be repeated. **Documentation** An inventory of occupational health risks shall be prepared. All measurements shall be well documented, stating measurement point, time, result, special observations, etc. The reporting shall be done on a special form. All reports shall be signed by the person conducting the measurements and by the production representative responsible. The production representative responsible shall communicate the measurement results to the personnel concerned. The report shall be distributed to the: • production manager; • department safety representative; • main safety representative. The report and the measurement protocol shall be filed according to a separate instruction.

Chemicals Handling Inc., SHE Management Procedures **Procedure for Occupational Health**			
Reg. No: SHE-9.4	Revision: 1	Valid from:	Page 2 of 2

Principles and methods (cont'd)	**Follow-up** Measurement results shall be summarized and analysed for trends at least once per year. Follow-up and reporting shall be made in writing in the quarterly report and the yearly report, and also regularly in department meetings, management group meetings, board meetings and SHE committee meetings.
Responsibility	The Managing Director has overall responsibility for ensuring that a programme to control risks to occupational health is in place. Department managers are responsible for the inventory, measurement programmes, measurements, evaluations, reporting and actions. The SHE committee deals with the results and decides on the extent and level of the programmes and actions.
References	This procedure refers to the following SHE procedures: SHE POLICY SHE OBJECTIVES AND ACTION PLANS MANAGEMENT REVIEW SHE LEGISLATION SHE PERMITS SHE EFFECTS/IMPACTS SHE REPORTING HEALTH CARE TRAINING INTERNAL COMMUNICATION EXTERNAL COMMUNICATION INSTRUCTIONS PLANT INTEGRITY AND MAINTENANCE WORK PERMITS CONTRACTORS HANDLING OF CHEMICAL PRODUCTS HANDLING OF WASTES MANAGEMENT OF CHANGE INVESTMENT PROJECTS MANUFACTURING METHODS/R&D WORK/SCALE-UP SHE AUDITS RISK ANALYSIS SHE ROUNDS ACCIDENTS, INCIDENTS AND DISTURBANCES PRODUCT CONTROL

Chemicals Handling Inc., SHE Management Procedures **Procedure for Environmental Control**				
Reg. No: SHE-9.5	Revision: 1	Valid from:		Page 1 of 3
Approved by:	Date:		Issued by:	
Distribution:				
Objective	The objective of this procedure is to describe Chemicals Handling Inc.'s systematic control of emissions to the environment.			
Scope	This procedure comprises the control of emissions of a continuous or intermittent character. Waste handling is regulated in a special procedure. Failures of equipment resulting in releases are regulated in another procedure.			

Chemicals Handling Inc., SHE Management Procedures **Procedure for Environmental Control**			
Reg. No: SHE-9.5	Revision: 1	Valid from:	Page 2 of 3

Principles and methods (cont'd)	**Action in case of deviation** When a measurement indicates a deviation from the permit values or internal target values, reporting shall be made to the department manager and: • to the SHE function when exceeding an internal target value; • to the SHE function for formal reporting to the relevant authority when exceeding a permit value. The procedure for ACCIDENTS, INCIDENTS AND DISTURBANCES, SHE-11.2, is applicable.
Responsibility	The site manager/managing director has overall responsibility for environmental control and for ensuring that the company fulfils the obligations in the permits. Responsibility for controlling emissions to the environment lies with the department managers. The SHE function is responsible for internal records, summaries and reporting to the authorities.
References	This procedure refers to the following SHE procedures: SHE POLICY SHE REPORTING MANAGEMENT REVIEW HANDLING OF WASTES SHE LEGISLATION SHE AUDITS SHE PERMITS ACCIDENTS, INCIDENTS AND DISTURBANCES SHE EFFECTS/IMPACTS

Chemicals Handling Inc., SHE Management Procedures
Procedure for Environmental Control
Attachment 1 – Example of emission measurements

| Reg. No: SHE-9.5 | | Revision: 1 | Valid From | | | | Page 3 of 3 Attachment 1 | |

Emission source	Point of measurement	Measurement system	Variable	Frequency	Permit value	Target value	Calibration	Comments
Process ventilation		Manual sampling/ laboratory analysis	Dust	1/month	$x\,mg/m^3$	0.5 x	1/quarter	
		Continuous analysis	Solvents	Continuous	$y\,mg/m^3$	0.8 y	1/month	
Process water		Manual sampling/ laboratory analysis	Organic material	1/week	$z\,mg/m^3$	0.7 z	1/month	
Emission from waste water treatment plant		24 hour sample	Organic material	1/day	$v\,mg/m^3$	0.8 v		
		Manual sampling	Suspended material	1/week	$w\,mg/m^3$	0.6 w		

Chemicals Handling Inc., SHE Management Procedures **Procedure for Emergency Response**				
Reg. No: SHE-10.1		Revision: 1	Valid from:	Page 1 of 2
Approved by:		Date:		Issued by:
Distribution:				

Objective	The objective of this procedure is to give principles and guidelines for Chemicals Handling Inc.'s emergency response plan. There shall be a special detailed emergency plan based on scenarios identified in risk analyses.
Scope	This procedure applies to all activities within the industrial site of Chemicals Handling Inc. Preparedness for transport accidents is also included. 'Emergencies' is taken to mean accidents or serious precursors to accidents with possible consequences for employees, the environment or property, such as accidental releases to the atmosphere, water or soil, fires and explosions.

Principles and methods	The objective of the emergency plan is that the company itself, in co-operation with the community rescue services (fire brigade, etc.), shall be able to deal efficiently with emergency situations and limit damages.
	   The emergency plan shall work at all times of the day and the year. Therefore a special emergency organization and a special room with equipment for leading the activities are needed.
	
	Emergency organization
	An emergency plan shall be prepared which clearly defines the organization for dealing with an emergency and the responsibility for specific actions. In preparing the plan the following shall be considered:
	
	• raising the alarm;
	• informing local emergency services – fire, police, ambulance;
	• provision of safe areas including, if necessary, 'toxic refuges' for staff, marking of safe areas and evacuation routes;
	• evacuation and roll call of staff;
	• treatment of casualties;
	• immediate response and control of incident;
	• rendezvous with a briefing of local emergency services;
	• ongoing liaison with emergency services (these later points will usually be dealt with by a member of the emergency management team);
	• provision of information to neighbours;
	• provision of information to the local and national media;
	• minimization of long-term damage;
	• restitution.
	
	   In preparing the emergency plan the following should be considered:
	
	• formation of an 'emergency management team' comprising the site manager, production manager, SHE manager and personnel manager;
	• provision of an 'emergency centre' (see below).
	
	During non-office hours at least one member of the 'emergency management team' should be available 'on call' to attend the site during any emergency.
	
	Emergency centre
	The emergency centre shall be located in a safe place. In close vicinity of this room is another room containing equipment to be used in case of an emergency situation, for example:
	
	• additional (ex-directory) telephone line;
	• extra telephones to be connected to a special extension;
	• system drawings of special utility systems, sewer systems, etc.;
	• layout drawings and plot plans;

Chemicals Handling Inc., SHE Management Procedures **Procedure for Emergency Response**				
Reg. No: SHE-10.1		Revision: 1	Valid from:	Page 2 of 2

Principles and methods (cont'd)	• telephone lists for key functions and telephone books; • information about the properties of the chemicals used; • tape recorder (for recording the conversation in the emergency centre); • log-book; • inventory of emergency equipment and facilities; • copy of emergency plan. **Communication plan** There shall be 'ready-made' plans for how communication in an emergency shall be established and which type of information shall be given. **Action plan** In the detailed plan, the immediate actions to be taken in certain standardized emergency situations shall be outlined. **Emergency response for transport accidents** The detailed emergency plan shall contain a section about how transportation accidents including company products shall be handled. Chemicals Handling Inc. shall be able to assist with certain expertise in such situations. Chemicals Handling Inc. is associated with the chemical industries ERC (Emergency Response Centre) for 24-hour coverage. **Exercises/drills** Exercises/drills shall be carried out regularly. The management shall have one exercise per year. In addition, exercises with the participation of the community rescue services shall be held at agreed intervals. **Training** All employees shall be trained in the emergency plan. Contractors shall also have adequate information about the plan to be able to act correctly in a real emergency situation.
Responsibility	The site manager is responsible for establishing an emergency plan with an emergency organization, and securing a working emergency preparedness in other respects. The SHE function shall do much of the practical work. The site manager is responsible for initiating emergency drills.
References	This procedure refers to the following SHE procedures: SHE POLICY CONTRACTORS SHE LEGISLATION RISK ANALYSIS TRAINING ACCIDENTS, INCIDENTS AND DISTURBANCES INTERNAL COMMUNICATION FIRE PROTECTION EXTERNAL COMMUNICATION TRANSPORT WORK PERMITS

Chemicals Handling Inc., SHE Management Procedures				
Procedure for Fire Protection				
Reg. No: SHE-10.2	Revision: 1	Valid from:		Page 1 of 4
Approved by:	Date:		Issued by:	
Distribution:				

Objective	The objective of this procedure is to give principles and guidelines for, and the most important contents of, Chemicals Handling Inc.'s fire protection plan, to ensure that there is a detailed plan: • to reduce fire risks; • to ensure that systems and equipment are available to minimize the effects of fire. Measures to control fire-fighting are covered in the procedure for EMERGENCY RESPONSE, SHE-10.1.
Scope	This procedure applies to all activities within the industrial site of Chemicals Handling Inc.

Principles and methods	Chemicals Handling Inc.'s fire protection plan is primarily intended to be preventive. The company resources for intervention in case of fire are limited to the initial acute phase, until external assistance arrives.
	    As part of the assessment of SHE effects, the department managers shall ensure that special attention is given to fire risks, their prevention and mitigation.
	    The available fire protection equipment and other resources shall be based on a systematic review of possible fire scenarios. The scale of resources shall take into consideration both the company resources for the initial intervention and the external resources for totally combating all plausible fire scenarios (Attachment 1 gives an example).
	
	 Fire protection
	 The following important fire protection principles apply:
	
	 • all rooms and spaces shall be kept clear of combustible material and rubbish;
	 • roads and access roads shall always be kept clear and passable;
	 • fire protection equipment shall always be in its proper place and may not be blocked or rendered unusable. Fire protection equipment may only be used for its intended purpose;
	 • all personnel shall be well trained and safety conscious;
	 • company general procedures concerning fire avoidance (e.g., at hot work) shall be followed up.
	
	 Fire protection equipment
	 There should be the necessary equipment on site for an initial efficient extinguishing intervention – fixed installed fire protection equipment as well as mobile fire extinguishing material located at special places. All equipment shall be regularly controlled and tested.
	    Attachment 1 is an example of a register of available extinguishing equipment, indicating the requirements of control, etc.
	
	 Alarms
	 Alarms, both push-button and automatic, shall be installed to:
	
	 • warn all employees of fire;
	 • notify the local fire brigade.
	
	 Schedules shall be in place for the testing of all alarms.
	
	 Training
	 All employees shall receive basic fire protection training with refresher courses every year. Contractors shall also have adequate information to be able to act correctly in a real emergency situation.

Chemicals Handling Inc., SHE Management Procedures **Procedure for Fire Protection**				
Reg. No: SHE-10.2		Revision: 1	Valid from:	Page 2 of 4
Principles and methods (cont'd)	The employees who constitute the internal Chemicals Handling Inc. intervention force shall receive rigorous training through external courses and yearly qualified refresher training.			
Responsibility	The site manager has overall responsibility for fire protection and the fire protection plan. The SHE manager is responsible for maintaining a good standard of preventive fire protection, ensuring that training is taking place and that exercises are carried out according to this procedure.			
References	This procedure refers to the following SHE procedures: SHE POLICY INSTRUCTIONS ORGANIZATION INVESTMENT PROJECTS SHE LEGISLATION SHE AUDITS TRAINING RISK ANALYSIS INTERNAL COMMUNICATION SHE ROUNDS EXTERNAL COMMUNICATION EMERGENCY RESPONSE GENERAL SITE RULES ACCIDENTS, INCIDENTS AND DISTURBANCES			

Chemicals Handling Inc., SHE Management Procedures
Procedure for Fire Protection
Attachment 1 – Fire protection plan: example of fire protection equipment

Reg. No: SHE-10.2	Revision: 1	Valid From	Page 3 of 4
			Attachment 1

Location	Equipment	Controls	Comments
The site, central	Fire water system	Capacity test, once per year.	
The site, general	Fire monitors	Operate all valves once per year, rolling schedule. Check all outside fire monitors against freezing risk once per week during the winter period.	
	Hose stations	Check once per month.	
	Hand extinguishers	Check once per year, rolling schedule.	
Production building	Sprinkling systems	Yearly control by an authorized company. Test run of pumps. Test of automation systems once per quarter.	
	Foam units	Check once per quarter.	
Tank storage	Foam sprinkling units for solvent tanks	Check once per quarter.	
Switch yard	Carbon dioxide unit	Check once per quarter.	
Office	Sprinkling units	Yearly control by an authorized company. Test run of pumps. Test of automation systems once per quarter.	

NB: Checks and controls shall be documented.

Chemicals Handling Inc., SHE Management Procedures
Procedure for Fire Protection
Attachment 2 – Fire protection plan: example of possible fire scenarios

Reg. No: SHE-10.2		Revision: 1	Valid from:		Page 4 of 4
					Attachment 2

Fire location	**Possible cause**	**Intervention actions**	**Fire equipment**	**Comments**
Fire in production building • production vessel	Leakage. Equipment not inerted; static electricity as ignition source.	Stop the unit. Alarm the fire brigade. Stop the feed; stop the leakage. Extinguish the fire.	Activate the foam sprinkling system.	Use protection mask/air bottles.
Fire in production building • filling equipment	Over-filling, leakage; static electricity as ignition source.	Stop the unit. Alarm the fire brigade. Stop the feed; stop the leakage. Extinguish the fire.	Hand extinguishers (foam). Activate the sprinkling system.	
Fire in tank storage				
Fire in switch yard				
Fire in office room				
Fire in container for packing				

Chemicals Handling Inc., SHE Management Procedures **Procedure for Actions in the case of Accidents, Incidents and Disturbances**				
Reg. No: SHE-11.1	Revision: 1	Valid from:		Page 1 of 2
Approved by:	Date:		Issued by:	
Distribution:				

Objective	The objective of this procedure is to ensure that all personnel are aware of the actions required in the event of an accident, incident or disturbance. Through the implementation of correct actions a situation can be controlled or mitigated, thus reducing the event's impact on personnel, the public or environment.
Scope	This procedure applies to all activities within Chemicals Handling Inc., all employees and all contracted and visiting personnel. It is also valid for the transport of company products. All accidents involving personnel injuries, fires, explosions, failures of equipment, releases of gas or liquid, environmental disturbances or forerunners to such accidents (incidents or 'near-misses') should be included.
Principles and methods	**Immediate actions** The immediate supervisor/manager shall as soon as possible go to the event and take the necessary immediate actions to make the situation safe and limit further damage. This shall be documented in the incident report (see SHE-11.2). A clear and concise action list/emergency plan should be produced by the company which is specific to the hazards and scenarios at the plant/unit (see SHE-10.1) Fatality, serious personnel injury or events which could lead to such consequences should be reported immediately to the relevant authorities. The relevant department manager is responsible for this. In case of environmental disturbances, the authorities for this shall be contacted and informed. The SHE function is responsible for this. Personnel injury which leads to absence from work or work-related injury shall be reported to the relevant authority (normally on a special form). The relevant department manager reports to the personnel function, which reports further to the authority. **Contact with family member in case of personnel injury** In case of personnel injury and transport to hospital, it is the responsibility of the relevant department manager to contact family member(s). If the injury is serious, the personnel manager and the site manager shall be informed for further action.
Responsibility	The relevant department manager has overall responsibility for the functioning of this procedure. Responsibility is also regulated in the above procedure.
References	This procedure refers to the following SHE procedures: SHE POLICY EXTERNAL COMMUNICATION MANAGEMENT REVIEW SHE AUDITS SHE LEGISLATION SHE ROUNDS SHE REPORTING OCCUPATIONAL HEALTH HEALTH CARE ENVIRONMENTAL CONTROL TRAINING EMERGENCY RESPONSE INTERNAL COMMUNICATION TRANSPORT

Chemicals Handling Inc., SHE Management Procedures				
Procedure for Investigation and Reporting of Accidents, Incidents and Disturbances				
Reg. No: SHE-11.2	Revision: 1	Valid from:		Page 1 of 2
Approved by:	Date:		Issued by:	
Distribution:				

Objective	The objective of this procedure is to ensure that the maximum possible lessons are learnt from accidents, incidents and disturbances. By preventive and corrective actions, similar events should be avoided and a continual increase in safety, health and environmental standards should be achieved.
Scope	This procedure applies to all activities within Chemicals Handling Inc., all employees and all contracted and visiting personnel. It is also valid for the transport of company products. All accidents involving personnel injuries, fires, explosions, failures of equipment, releases of gas or liquid, environmental disturbances or forerunners to such accidents (incidents or 'near-misses') should be reported. Deviations from procedures which could lead to a hazardous situation are also considered as incidents in this context. There is no differentiation made between accidents or incidents with personnel injuries and purely technical incidents; all events shall be reported in the same way. Work-related injuries and illness shall also be reported in this system.
Principles and methods	**Reporting** An incident report shall be written for every accident, incident or disturbance which occurs within Chemicals Handling Inc. or with company products during transport. The next higher manager in the organization is responsible for reporting each event which will be done in co-operation with the employee concerned. The incident report shall be written on a special form (to be designed by the company) on the same day or shift as the event occurred. The report shall be signed by the immediate supervisor/manager, the employee and the relevant safety representative. The original shall be sent to the respective line manager. Copies shall be sent to the relevant department manager, SHE function, the relevant safety representative and the main safety representative, and finally posted on the notice board of the department. **Further actions** As soon as possible after the incident has occurred, the department manager shall establish the causes of the event and decide on possible actions. Further investigation of the event shall be initiated when necessary. This decision shall be documented in the incident report. The department manager is responsible for deciding the actions to be carried out to determine the root caused and actions to prevent a recurrence. **Special actions in case of serious accident** A special investigation shall be carried out for serious accidents. Examples of such events are serious personnel injury, serious fire or explosion, major failure of equipment, which have affected or could affect the environment. The department manager or the site manager is responsible for ensuring that the special investigation is carried out to determine the root causes and actions to prevent a recurrence. **Detailed follow-up and analysis** When the investigation and the agreed actions have been carried out, the original copy of the report shall be sent to the SHE function for follow-up and filing. An analysis of the incident shall be carried out to decide the direct and root causes. Regular summaries shall be made of all the damages and incidents, based on the type of incidents and the causes. The summaries shall be presented to the SHE committee, which shall recommend further action programmes when relevant, which in turn shall be agreed by the management team.

Chemicals Handling Inc., SHE Management Procedures			
Procedure for Investigation and Reporting of Accidents, Incidents and Disturbances			
Reg. No: SHE-11.2	Revision: 1	Valid from:	Page 2 of 2

Principles and methods (cont'd)	**Documentation** The incident (deviation) report (to be designed by the company) shall, among other things, contain: • a description and data of the event including causes, effects and immediate actions taken; • decisions of the department manager (plan for corrective actions); • references to investigations and/or work orders; • follow-up that agreed actions are carried out and have given desired effects. See the incident notification form in 'Benchmarking on EPSC member company incident reporting systems', www.epsc.org/papers.htm#paper1 **WHAT IS AN INCIDENT?** As well as the obvious cases of incidents, such as a release, events of the following type should also be regarded as incidents: • bolt dropped from high elevation; • safety valve appeared to be blocked at the yearly inspection; • vital alarm function did not work; • contractor worked without valid work permit. **It is better to report one incident too many than one too few!**
Responsibility	The relevant department manager has overall responsibility for the functioning of this procedure. Responsibility is also regulated in the above procedure.
References	This procedure refers to the following SHE procedures: SHE POLICY EXTERNAL COMMUNICATION MANAGEMENT REVIEW SHE AUDITS SHE LEGISLATION SHE ROUNDS SHE REPORTING OCCUPATIONAL HEALTH HEALTH CARE ENVIRONMENTAL CONTROL TRAINING EMERGENCY RESPONSE INTERNAL COMMUNICATION TRANSPORT

Chemicals Handling Inc., SHE Management Procedures **Procedure for Transport**				
Reg. No: SHE-12.1	Revision: 1	Valid from:		Page 1 of 2
Approved by:	Date:		Issued by:	
Distribution:				

Objective	The objective of this procedure is to provide guidance on transport activities so that they are performed in a way which reduces the safety, health and environmental risks so far as reasonably practicable.
Scope	This procedure applies to all transport activities within Chemicals Handling Inc.'s industrial site and transport of the company's products outside the company. It also includes the use of vehicles within the industrial site.
Principles and methods	Chemicals Handling Inc. shall evaluate the SHE aspects for all types of transportation performed for the company (by own or external vehicles) and shall decide which type shall be used, taking into account other aspects, such as economics. **Transport of the company products** The following issues shall apply when transporting company products: • All formal legislative requirements shall be fulfilled, especially those concerning the transport of dangerous goods. • Only vehicles approved for the purpose shall be used. • Transporters shall only use officially approved/recommended route selection. • The choice of transporter shall be made according to the section 'External transport' below. • Chemicals Handling Inc. shall make a special check that prescribed transport documents are always present during transportation, for example declaration of goods, transport cards. • Chemicals Handling Inc. shall check that dispatched goods have the prescribed labelling on the packaging and/or the vehicle. • Safety equipment prescribed by the transport authority shall be present during transportation. • In case of a transport accident involving company products, the company shall be prepared to assist by providing information and, in certain cases, active assistance at the actual place. The company is associated with the national ERC (Emergency Response Centre) (see procedure for EMERGENCY RESPONSE, SHE-10.1). **Transport using Chemicals Handling Inc.'s own vehicles** Chemicals Handling Inc. has vehicles (pick-up, truck and fork-lift) for transportation within the industrial site and in the vicinity of the company. Only personnel with approved training for the respective vehicles are allowed to drive them. This shall be controlled by the responsible supervisor/manager when lending a vehicle. The site maintenance manager shall ensure that all vehicles (including those used only on-site) are subject to regular maintenance and inspection to agreed standards. **External transport** In appointing an operator to transport products off-site the following factors shall be taken into account before the award of contract: • SHE policy of the transporter; • standard of equipment, vehicles, hoses, couplings, etc.; • training of drivers;

Chemicals Handling Inc., SHE Management Procedures **Procedure for Transport**			
Reg. No: SHE-12.1	Revision: 1	Valid from:	Page 2 of 2

Principles and methods (cont'd)	• experience and safety record of the transporter; • environmental impact of vehicles; • arrangements to audit the above. **Loading and unloading** The department manager is responsible for ensuring that suitable procedures and instructions are in place to ensure that: • the materials delivered to site are correct; • loading or unloading is undertaken in a way which is safe and will not damage the environment. See *Procedures for Off-loading Products into Bulk Stores at Plants and Terminals*, published May 1999 by the Chemical Industries Association (e-mail: publications@cia.org.uk).
Responsibility	The SHE manager is responsible for the company fulfilling legislative requirements for the handling of dangerous goods. The production supervisor is responsible for ensuring that vehicles leaving the company with dangerous goods will fulfil the requirements of this procedure. Spot tests of vehicles shall be made. The purchasing person is responsible for ensuring that all hired vehicles fulfil the requirements of this procedure. The purchasing manager shall arrange for audits of external transport operations.
References	This procedure refers to the following SHE procedures: SHE POLICY HANDLING OF CHEMICAL PRODUCTS ORGANIZATION HANDLING OF WASTES SHE LEGISLATION SHE AUDITS SHE PERMITS RISK ANALYSIS SHE EFFECTS/IMPACTS SHE ROUNDS TRAINING EMERGENCY RESPONSE PURCHASING ACCIDENTS, INCIDENTS AND DISTURBANCES INSTRUCTIONS PRODUCT CONTROL CONTRACTORS

Chemicals Handling Inc., SHE Management Procedures **Procedure for Product Control**				
Reg. No: SHE-13.1	Revision: 1	Valid from:		Page 1 of 2
Approved by:		Date:	Issued by:	
Distribution:				

Objective	The objective of this procedure is to clarify the requirements and the rules for handling product control issues within Chemicals Handling Inc.
Scope	This procedure applies to all chemical products handled by Chemicals Handling Inc., both produced by the company and imported.
Principles and methods	Commercial handling of chemicals is regulated by a number of directives and other legislation. As a basic rule these shall be followed (see procedure for SHE LEGISLATION, SHE-3.1). Chemicals Handling Inc. shall have the required permits for handling certain chemicals (see procedure for SHE PERMITS, SHE-3.2). No chemicals may be handled before review, approval and registration have been completed. As a rule, all chemicals shall have an approved material data sheet (SHE) before handling is allowed. The SHE aspects, including safety, human toxicological and environmental toxicological aspects, of all products which are produced by or enter into the company shall be satisfactorily evaluated. **CHEMICALS HANDLING INC.'S PRODUCTS** **Investigations** At selection and production of new products, the SHE aspects shall be investigated and alternative products and manufacturing routes shall be evaluated before a decision is made (according to the procedure for MANUFACTURING METHODS/R&D WORK/SCALE-UP, SHE-8.5). Simplified life cycle analyses shall be carried out for totally new products and further simplified analyses carried out for modified products. Products shall, as far as possible, be designed for recycle, reuse and recovery. **Information** Material safety data sheets (SHE) and handling instructions shall be available for all products leaving the company. The company shall keep an updated register of all customers. Internal safety and environmental data sheets shall be at hand when working in the laboratory and the pilot plant. All products shall be properly labelled according to the applicable rules and regulations. All Chemicals Handling Inc. products shall be listed in the product register of the relevant authority as soon as possible. The production manager is responsible for the classification and labelling of products manufactured within the company. **Training** Chemicals Handling Inc. will provide customers buying large quantities of company products (and also certain others such as transporters) with special training, on the SHE aspects of the products. **Packaging** Packaging for company products shall preferably be designed in such a way that it can normally be reused. Alternatively it shall be possible to dispose of the packaging in a safe and environmentally acceptable way through combustion or deposition. Chemicals Handling Inc. shall manage to take care of certain larger packaging. Chemicals Handling Inc. is listed on the national register for packaging.

Chemicals Handling Inc., SHE Management Procedures **Procedure for Product Control**			
Reg. No: SHE-13.1	Revision: 1	Valid from:	Page 2 of 2

Principles and methods (cont'd)	**OTHER CHEMICAL PRODUCTS** **Purchase of chemicals** See procedure for PURCHASING, SHE-6.1. **Investigation** Before a chemical product can be introduced into the company, its SHE properties shall be investigated. This investigation shall be made by the SHE function or by external expertise engaged by the SHE function. **Chemicals register** The company shall maintain an up-to-date register of the dangerous substances that are handled. The register shall be kept open and available to all employees and shall refer to the SHE data sheets. **Safety, health and environmental information** The manufacturer or importer of a dangerous substance is obliged to supply detailed risk and safety (SHE) information in the form of material safety (SHE) data sheets. The purchasing function is responsible for ensuring that data sheets are obtained before the product arrives at the company. The SHE function scrutinizes the data sheet and decides whether it can be used in this form, otherwise an internal data sheet must be produced. Data sheets shall be distributed to the users of the product by the SHE function. **Labelling** When the chemicals arrive at the company, the packaging shall be checked to make sure that it is correctly labelled with clear labels giving the name of the substance, material number, applicable risk label, information on possible specific risks, etc. These shall be labelled in the same way when dispensing into in-house packaging. The production manager is responsible for the correct labelling of materials.
Responsibility	The production manager is responsible for ensuring that all products leaving the company are provided with the information as specified above. The production manager is responsible for following the regulations concerning internally handled chemicals.
References	This procedure refers to the following SHE procedures: SHE POLICY · INSTRUCTIONS MANAGEMENT REVIEW · HANDLING OF CHEMICAL PRODUCTS SHE LEGISLATION · HANDLING OF WASTES SHE EFFECTS/IMPACTS · MANUFACTURING METHODS/R&D WORK/SCALE-UP SHE REPORTING · SHE AUDITS TRAINING · OCCUPATIONAL HEALTH PURCHASING · TRANSPORT

Further reading

The following titles have been chosen to provide additional information which readers of this guide may find useful.

0.   SHE management system, general

EPSC, 1994, *Safety Management Systems: Sharing experiences in process safety*, pages 5–8 (IChemE)

IChemE Training Package 032, *Safety Management Systems*, 1998

IChemE Training Package 017, *Managing for Safety*, 1991

1.   SHE policy

EPSC, 1994, *Safety Management Systems*, page 10 (IChemE)

4.   General SHE information

Wells, G., 1997, *Major Hazards and Their Management* (IChemE)

5.2   Training
IChemE training packages and publications in general

7.   Operations and maintenance

IChemE Training Package 035, *Risk Assessment Techniques*, 2000

IChemE Training Package 033, *Safer Maintenance*, 1999

IChemE Training Package 025, *Modifications: The Management of Change*, 1994

EPSC, 2000, *Safety Integrity: The Implications of IEC 61508 and Other Standards for the Process Industries* (IChemE)

Barton, J. and Rogers R. (eds), 1997, *Chemical Reaction Hazards: A guide to safety*, 2nd edition (IChemE).

Pitblado, R. and Turney, R., 1996, *Risk Assessment in the Process Industries*, 2nd edition (IChemE)

9.   Auditing and inspection

EPSC, 1994, *Safety Management Systems: Sharing experiences in process safety*, page 13 (IChemE)

IChemE Training Package 031, *Safety Auditing*, 1998

Wells, G., 1996, *Hazard Identification and Risk Assessment* (IChemE)

Association of Swedish Chemical Industries, 1996, *SHE-AUDIT, A Guideline for Internal Auditing of SAFETY/HEALTH/ENVIRONMENT*

11.   Accidents, incidents and disturbances

IChemE Training Package 026, *Incident Reporting: investigation and analysis*, 1995

Appendix 1 – Reference key to ISO 14001: 1996 (EMAS)

A written environmental report should be included for comparison with EMAS. ISO 14001 is approved as an environmental management system for EMAS. (The environmental report may be found under 4.2 SHE reporting.)

ISO 14001 (Environmental management system requirements)	SHE management system
4.1 General requirements	SHE management system
4.2 Environmental policy	1. SHE policy
4.3.1 Environmental aspects	SHE management system, particularly: 4.1 SHE effects/impacts 7.10 Management of change 8.2 Investment projects 8.3 Manufacturing methods/R&D work/scale-up 9.1 SHE audits 9.2 Risk analysis
4.3.2 Legal and other requirements	3. Legislation and permits
4.3.3 Objectives and targets	2.2 SHE objectives and action plans
4.3.4 Environmental management programme(s)	2.2 SHE objectives and action plans 8.2 Investment projects 8.4 Manufacturing methods/R&D work/scale-up
4.4.1 Structure and responsibility	2.1 Organization
4.4.2 Training, awareness and competence	5.2 Training
4.4.3 Communication	5.3 Internal communication 5.4 External communication
4.4.4 Environmental management system documentation	0. SHE management system, general
4.4.5 Document control	0. SHE management system, general
4.4.6 Operational control	4.1 SHE effects/impacts 5.2 Training 6.1 Purchasing 7. Operations and maintenance 8. Engineering and projects 12. Transport 13. Product control
4.4.7 Emergency preparedness and response	10.1 Emergency response
4.5.1 Monitoring and measurement	9.1 SHE audits 9.2 Risk analysis 9.3 SHE rounds 9.5 Environmental control

Continued overleaf

ISO 14001 (Environmental management system requirements)	SHE management system	
4.5.2 Nonconformance and corrective and preventive action	4.2	SHE reporting
	9.5	Environmental control
	11.	Accidents, incidents and disturbances
4.5.3 Records	2.3	Management review
	4.1	SHE effects/impacts
	4.2	SHE reporting
	5.2	Training
	8.	Engineering and projects
	9.1	SHE audits
	9.2	Risk analysis
	9.3	SHE rounds
	9.5	Environmental control
4.5.4 Environmental management system audit	9.1	SHE audits
4.6 Management review	2.3	Management review

Appendix 2 – Reference key to BS 8800: 1996

BS 8800 (Occupational health and safety management systems)	SHE management system
4.0 Introduction 4.0.1 General	SHE management system
4.0.2 Initial status review	4.1 SHE effects/impacts
4.1 OH&S policy	1.0 SHE policy
4.2 Planning 4.2.1 General	4.1 SHE effects/impacts 9.2 Risk analysis 9.4 Occupational health
4.2.2 Risk assessment	4.1 SHE effects/impacts 9.2 Risk analysis 9.4 Occupational health
4.2.3 Legal and other requirements	3.1 SHE legislation 3.2 SHE permits
4.2.4 OH&S management arrangements	2.1 Organization 2.2 SHE objectives and action plans + majority of other procedures
4.3 Implementation and operation 4.3.1 Structure and responsibility	2.1 Organization
4.3.2 Training, awareness and competence	5.2 Training
4.3.3 Communications	5.3 Internal communication 5.4 External communication
4.3.4 OH&S management system documentation	0. SHE management system, general
4.3.5 Document control	0. SHE management system, general
4.3.6 Operational control	4.1 SHE effects/impacts 5.2 Training 6.1 Purchasing 7. Operations and maintenance 8. Engineering and projects 12. Transport 13. Product control
4.3.7 Emergency preparedness and response	10.1 Emergency response 10.2 Fire protection

Continued overleaf

BS 8800 (Occupational health and safety management systems)	SHE management system
4.4 Checking and corrective action 4.4.1 Monitoring and measurement	4.2 SHE reporting 5.1 Health care 7.1 General site rules 7.2 Instructions 7.3 Plant integrity and maintenance 7.4 Work permits 7.5 Contractors 7.6 Handling of chemical products 9.1 SHE audits 9.2 Risk analysis 9.3 SHE rounds 9.4 Occupational health 11. Accidents, incidents and disturbances 13. Product control
4.4.2 Corrective action	11. Accidents, incidents and disturbances
4.4.3 Records	4.2 SHE reporting 5.1 Health care 7.6 Handling of chemical products 9.2 Risk analysis 9.3 SHE rounds 9.4 Occupational health 11. Accidents, incidents and disturbances 12. Product control
4.4.4 Audit	2.3 Management review 9.1 SHE audit
4.5 Management review	2.3 Management review